PRAISE FOR NEUROTEACH

"Education can sometimes seem bedeviled by myths and folklore not based on anything other than trendy-sounding pseudoscience and gut instinct. The problem is that much of what teachers and students take for granted is just plain wrong. While cognitive psychology is playing an increasingly important role in how teachers understand their craft and how students can best learn, neuroscience has, for the most part, remained the realm of quacks and snake-oil salesmen. *Neuroteach* combines important lessons from both disciplines to draw useful, practical conclusions that will clarify not just what teachers should do in the classroom, but to help them better understand the complexities of the science of learning." —**David Didau (@LearningSpy), author of *What If Everything You Knew About Education Was Wrong?* and *The Secret of Literacy***

"Integrating educational neuroscience into practice is best accomplished when educators become the neuroscience research translators, taking valid research from the lab to the classroom. *Neuroteach* is not simply a how-to list of practical things to do based on neuroscience research, it is a guide for educators to become independent neuroeducators through using its models and frameworks. It also provides practical guidance about using theoretically sound, research-based principles in the design of their schools, classrooms, and work with individual students in their classrooms.

"I share and applaud the authors' goal that the neuroscience of learning be incorporated into all teacher-training programs and sustained with ongoing professional development and professional learning communities at schools. This training and development is both relevant and timely as neuroscience enters a new accelerated phase of revealing how the brain learns through brain mapping, enhanced fMRI, and diffusion tensor imaging, which reveals networks throughout the brain that are activated together in different types of information processing.

"*Neuroteach* unravels the conceptual and practical as it guides educators in two critically needed components as they become neuroeducators: how to evaluate neuroscience research quality and then synthesize and translate it into practical applications. This guidance includes the criteria with which to judge the validity of any research and claims (avoiding neuromyths) and in how to translate the fruits of neuroscience research into practical applications to curriculum, instruction, and assessment." —**Judy Willis (@judywillis), author of *Research Strategies to Ignite Student Learning: Insights from a Neurologist and Classroom Teacher***

"I applaud Whitman and Kelleher for unapologetically releasing teachers from being the 'fish inside the bowl' of education research. No longer merely the subjects of research, *Neuroteach* lays out a framework for teachers as the gatekeepers of research findings that are relevant to classroom practice. While highlighting classroom teachers as the experts on learning, they underscore the importance of utilizing the strongest evidence in mind, brain, and education science to inform one's practice. As teachers writing from the lived experience of an effective model for school and university partnerships, Kelleher and Whitman synthesize and translate research into next-day practice." —**Dr. Vanessa Rodriguez (@teachingbrain), teacher, researcher, and author of *The Teaching Brain: An Evolutionary Trait at the Heart of Education***

"Teaching without an understanding of modern brain science is like practicing medicine without an understanding of genomics. In this integration of learning, classroom practice, and cutting-edge research on how the brain actually works, Whitman and Kelleher have written a book that belongs on the shelf and in the practical working sphere of every K–12 educator today. To my knowledge there is no school in America as deeply embedded in the practical study and application of brain research and student performance as St. Andrew's, and no educators more completely qualified and effective at translating the remarkable impacts of neuroscience down from the 30,000-foot level to 'what can we do in class tomorrow morning' as these two educator-thought leaders."
—**Grant Lichtman (@GrantLichtman), author of #*EdJourney: A Roadmap to the Future of Education***

"Schools across the world are experiencing an unprecedented opportunity to inform their classroom practice with cognitive science, neuroscience, and behavioural psychology. The intersection of these sciences with education is starting to inform our understanding of how students learn, how they can understand their own strengths and weaknesses, how they can improve their attention, their motivation and their memory—all vitally important matters to students and to teachers.

"*Neuroteach* offers a much-needed bridge between good research and schools, guiding teachers and school leaders in *how* to translate the growing body of research in educational neuroscience into the design of their schools and classrooms, and offering models and frameworks for mobilizing that research in learning and teaching. The authors, classroom teachers who focus on the intersection of research and practice in their work at the Center for Transformative Teaching and Learning at St. Andrew's Episcopal School, call for teachers to take a fundamental step forward in professionalizing their practice and—crucially—give them the tools to do so. This exciting, well-informed and wonderfully practical book will be welcomed by teachers worldwide." —**Jonnie Noakes (@JonnieNoakes), Director, Tony Little Centre for Innovation and Research in Learning, Eton College, UK**

"Glenn Whitman and Dr. Ian Kelleher craft a truly empathetic vantage point of the challenges both practitioners and researchers face in the application of neuroeducational research to teaching practices. Teachers will find that *Neuroteach* maintains attention throughout with its honesty, brisk pacing, and useful suggestions for sound pedagogical practices. A guide suitable for all content areas and backgrounds, *Neuroteach* aims to provide interventions for all students regardless of their levels of academic attainment. The authors also champion research-informed instruction to provide students with a myriad of opportunities to cultivate critical knowledge and growth mindsets. The book format invites reader reflection and direct application of content. This level of respect for the process of teaching and learning is sustained throughout the guide, providing its distinct voice and strengthening the authors' call to incorporate neuroscience with pedagogy. 'All teachers can be neuroteachers' with strong advocates like Glenn Whitman and Dr. Ian Kelleher." —**Dr. Mariale Hardiman (@MarialeHardiman), Cofounder and Director, Johns Hopkins University School of Education's Neuro-Education Initiative and author of *The Brain-Targeted Teaching Model for 21st-Century Schools***

NEUROTEACH

Brain Science and the
Future of Education

Glenn Whitman and Ian Kelleher

ROWMAN & LITTLEFIELD
Lanham • Boulder • New York • London

Published by Rowman & Littlefield
A wholly owned subsidary of
The Rowman & Littlefield Publishing Group, Inc.
4501 Forbes Boulevard, Suite 200, Lanham, Maryland 20706
www.rowman.com

Unit A, Whitacre Mews, 26-34 Stannary Street, London SE11 4AB,
United Kingdom

British Library Cataloguing in Publication Information Available

Library of Congress Cataloging-in-Publication Data
Names: Whitman, Glenn, 1969– author. | Kelleher, Ian, 1970– author.
Title: Neuroteach : brain science and the future of education / Glenn
    Whitman and Ian Kelleher.
Description: Lanham, Maryland : Rowman & Littlefield, 2016. | Includes
    bibliographical references and index.
Identifiers: LCCN 2016014601 (print) | LCCN 2016022242 (ebook) | ISBN
    9781475825343 (cloth : alk. paper) | ISBN 9781475825350 (pbk. : alk.
    paper) | ISBN 9781475825367 (electronic)
Subjects: LCSH: Cognitive neuroscience. | Cognitive learning. | Education—
    Philosophy.
Classification: LCC QP360.5 .W47 2016 (print) | LCC QP360.5 (ebook) |
    DDC 612.8/233—dc23
LC record available at https://lccn.loc.gov/2016014601

∞™ The paper used in this publication meets the minimum requirements
of American National Standard for Information Sciences—Permanence of
Paper for Printed Library Materials, ANSI/NISO Z39.48-1992.

Printed in the United States of America

This book is dedicated to our children, Grace, Owen, Anwen, Tamsin, and those students, living and yet unborn, who all deserve teachers who know how the brain learns.

CONTENTS

ACKNOWLEDGMENTS

If there is a book that you want to read, but it hasn't been written yet,
you must be the one to write it.

—Toni Morrison

Our journey to writing *Neuroteach* began in 2007 when our school made
the decision to increase 100 percent of its teachers' knowledge of how the
brain works, learns, and changes through training and ongoing professional
development. Together with our colleagues and our university partners, we
have been on an incredible adventure that has brought important leaders in
the field of mind, brain, and education science, such as Dr. Kurt Fischer, Dr.
Jay Giedd, Dr. Mariale Hardiman, Dr. Christina Hinton, Dr. Luke Rinne,
Dr. Vanessa Rodriguez, and Dr. Judy Willis, to our school to tell us about
their research. We are thankful to them for sharing their expertise with us.

But having such great minds and leaders in the field of educational
neuroscience on campus did not by itself ensure that teachers would
transform their instructional design or work with each student in ways
informed by research. It took a growth-minded group of teachers, aided
by visionary school leaders and support from the Crimsonbridge Group,
to see that they can better challenge and support all of their students by
not only knowing the content of their subjects really well but also by better knowing research-informed pedagogical practices to help deliver that
content. So, our greatest thanks go out to the teachers and leaders of St.

Andrew's for their support, energy, and wisdom. This book would not be possible without you. We learned and innovated together to see what research was best suited for each of our grade levels, disciplines, and students. But we also recognized the larger public purpose that our school, the experience of its teachers, and its Center for Transformative Teaching and Learning could have. One of our inspirations for taking the time to write *Neuroteach* was that we would serve as a bridge between academic research and classroom practices. We hope this book equally serves individual teachers, while providing a model for school-wide integration of mind, brain, and education science.

Anyone who has written a school essay, college paper, graduate thesis, or book knows the importance of having good editing support. We are, therefore, grateful that our former colleague, Dr. Elizabeth Weber, delayed her disappearing into retirement long enough to wield her mighty red pen one more time to help bring greater consistency and continuity to each chapter of *Neuroteach*. Additional editing help was provided by the chief beneficiaries of this work: students, in this case the CTTL Student Research Fellows, who gladly critiqued the writing of their teachers. We would also like to thank our colleague Joe Phelan for making each of us look years younger in our author images at the end of this book. It is amazing what a good camera and lighting can do. And we are thankful for the artistry of Lisa Malveaux, who created the visual translation of many of our ideas. We are also eternally grateful for a long hike in Colorado that Glenn took with his wife, Debra, which led to this book's title, and for everyone in both our families for allowing us the wonderful indulgence to pause for a while, think, and write.

Finally, we are indebted to every reader—teacher, school leader, educational policy maker, parent, and student—who will take the time to read parts or all of *Neuroteach*. Like each of these stakeholders, we want every student, every day of their academic journey to receive a world-class, research-informed education. One slice of this must be recognizing that exceptional teachers and school leaders *have* to know how the brain learns and works. Hopefully, this book will begin an adventure for each of you, similar to the one we began with our colleagues in 2007. May yours be equally as rewarding.

INTRODUCTION

If you remember anything of this book, it will be because your brain is slightly different after you have finished reading it.

—Eric Kandel

Teachers are brain changers. Thus, it would seem obvious that an understanding of the brain—the organ of learning—would be critical to a teacher's readiness to work with students. A neuroteacher is therefore one who intentionally applies research from the field of mind, brain, and education science to his or her instructional design and work with every student. "We have learned more about how the brain actually learns in the past ten years than in the previous hundred" is a common phrase in this field, so with this in mind, should not all teachers be neuroteachers?

Unfortunately, in traditional public, public-charter, private, parochial, and home schools across the country, most teachers lack an understanding of how the brain receives, filters, consolidates, and applies learning for both the short and long term. Such knowledge is not yet part of the formal training all teachers receive prior to entering the classroom. The grand national and international debate talks about standards and what students should know, but this unprecedented moment in time, where strands of neuroscience, cognitive science, and behavioral psychology are beginning to intersect in ways that are giving rise to actionable practices, research-informed and classroom-tested calls for more.

There needs to be an equal commitment to how students learn that requires an understanding of the architecture and inner workings of the brain. *Neuroteach* was therefore written to help solve the problem teachers and school leaders have in knowing how to bring the growing body of educational neuroscience research into the design of their schools, classrooms, and work with each individual student. It is our hope that *Neuroteach* will help ensure that one day, every student—regardless of zip code or school type—will learn and develop with the guidance of a teacher who knows the research behind how his or her brain works and learns.

"Nurse, get on the Internet, go to Surgery911.com, scroll down and click on the 'Are You Lost?' icon."

Figure 0.1. Why "professional" professional development matters. Reprinted with permission of Jerry King.

Imagine for a moment a medical doctor coming into a consultation for your upcoming surgery and saying, "I have great instinct, passion, and love for your heart, but I have never formally studied it." Would you choose this doctor to conduct the procedure? Probably not. However, it is with great instinct, passion, and love, as well as idealism, that most teachers and school leaders arrive at and proceed through their careers as educators.

What teachers and school leaders lack is an understanding of the organ of learning—the brain—and how such understanding can help each student meet his or her peak potential while better understanding himself or herself as a learner.

There is no question that teaching is a challenging profession made increasingly difficult because of the complexity of the brain and the growing number of tools available to all teachers, school leaders, and students. But the most important tool that each student brings to learning is his or her brain. While students can claim they forgot their homework or notebook, they can never say that they forgot their brain. But even with their brain, there is no assurance that learning will happen.

Ultimately, what research shows is that there is no greater influence on student outcomes than teacher quality. As a result, the fundamental premise of this book is that educational neuroscience is the missing resource, a jeweler's loupe, in most educators' tool kit. It will verify some classic teaching practices but repudiate others—with evidence—and it will offer new strategies and structures. It is the next frontier for teacher and school-leadership training.[1]

Before going any further, we must address an issue common in emerging multidisciplinary fields—what to call what you are doing. "Educational neuroscience" and "mind, brain, and education science" are often used interchangeably. We favor mind, brain, and education science (MBE science) because it is sometimes used to suggest closer ties with classroom practice. For example, "MBE differs from previous efforts, such as educational neuroscience, in that it is focused on the problem of how we might bring findings from the learning sciences into the classroom. As such, MBE is placed squarely in the classroom, and works through engaging teachers as primary participants."[2] We are also deliberately avoiding the terms "brain-based" and "neuroeducation" as these are often associated with neuromyths.

For educational neuroscience to become part of each teacher's mindset—for this definition of MBE science to flourish—what is needed are models of translation of research and neuroscience integration frameworks. This book provides such models and frameworks, as well as specific ways teachers can apply research in the transdisciplinary field of MBE science.

Many successful teachers and school leaders will learn that some of their instincts about "best" teaching and learning practices are validated by educational neuroscience. However, many teachers and school leaders will also see how MBE science not only informs but also transforms their instructional practice and work with each student. In this journey, the role of research is important, and we agree with Dr. Christina Hinton, who says, "the Harvard

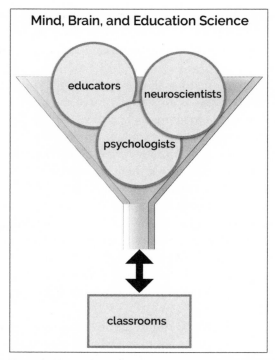

Figure 0.2.   What goes into mind, brain, and education science?

Ed School [has] been using the term 'research-informed' to emphasize that research should inform practice, but not dictate it. Teaching is both a science and an art, and this phrase leaves room for the 'art' of the practitioner."[3]

When thinking about the need for *Neuroteach*, we thought a lot about the question, "Why do individuals become and remain school teachers or leaders?" They make this commitment because they believe in the possibility that they can transform the lives of students, fellow faculty, or their school community. Initiating, maintaining, and deepening this commitment are the major challenges in education today. Can we meet these challenges with new scientific knowledge of the human brain? Can teacher training and ongoing professional development that are framed by innovative yet underutilized research in the field of MBE science enhance teacher quality and student outcomes? The answer is a resounding "yes!"

What we have found in the workshops we have led throughout the world is that the "best and the brightest" teachers and "rising star" school leaders, using "cutting edge" research, want to feel professionally challenged, to be directly connected to educational research, and to improve teaching and learning for every student in their schools. By placing into the hands of every teacher and school leader the growing body of research on the learning brain, the more effective they will be in their work with students and as leaders for change—thus making them more professionally fulfilled and increasing their interest in a lifelong career in education.

Research further validates this point. Enhanced teacher efficacy, a teacher's belief in his or her capacity to impact each student's learning, is strongly correlated to increased knowledge of the brain.[4] From our personal experience, the more our colleagues know about how the brain learns, works, and changes, the more differentiated their instructional practice becomes. Teachers quickly recognize that there are multiple pathways to the high academic expectations they establish for each one of their students. As teachers and school leaders become more MBE science–informed, they realize that the one-size-fits-all "industrial" model of education is not an equitable educational experience for all students.

People are screaming for better education. There is no greater opportunity to achieve this, yet no greater challenge, than integrating research from education, behavioral psychology, and cognitive science into the daily work of teachers and the thinking of school leaders. Imagine what is possible if we converged the expertise of scientists and teachers to uncover what teaching practices are confirmed by scientific study and thus should be continued, and what teaching practices that might intuitively make sense, either lack validation or have even been shown to be counterproductive.

What if we can replace the latter with methods that have been shown to work? What if we can spread this knowledge of a new definition of teaching excellence? By basing what we do on what is known, which sounds so gloriously simple and intuitive, we are on the tantalizing edge of the possibility of taking a major step forward in the professionalization of our practice.

There are four important groups of stakeholders in deepening the accurate use of educational neuroscience in the classroom: university researchers, teachers and school leaders, students, and parents. The "game changer" for the increased interest in connecting educational neuroscience with the daily classroom practices of teachers is the result of the growing awareness by university researchers that their work is only valuable if it gets into the hands of educators who might be able to scale up or broaden the application of an existing research study.

Historically, research in how the brain learns would be published in scholarly journals not usually read by precollegiate teachers and in technical language that is inaccessible to someone not trained in scientific fields. However, books by Dr. Mariale Hardiman, Dr. Tracey Tokuhama-

Esponosa, and Daniel Willingham and the scholarly journal *Mind, Brain, and Education* have bridged the gap between research and practice. Their work confirms that researchers and teachers need one another and that universities and schools should find more opportunities for collaborating and doing original research in the spirit of the research school idea put forth by John Dewey so many years ago.[5] In short, researchers and teachers each share authority for the ways educational neuroscience can inform, validate, and transform current pedagogical practice.

What about the other two groups of stakeholders: parents and students? There is a need for parents to have an easily accessible informational arsenal if they are to be effective allies to shift the great education debate to demand greater teacher quality and professional development. And the students? It is all about the students. It is all about helping them reach their maximum potential. We do this by aiding the other three groups of stakeholders, yes, but research also shows that putting mind, brain, and education science into the hands of young people in ways that are accessible to them actually empowers them to be more efficient, confident, and higher-achieving students.[6]

The challenges with integrating educational neuroscience into pedagogical practices emerge in the oversimplification of brain-based research and the neuromyths about how learning happens in the brain that might intuitively make sense but have no basis in research. There is also a lot of money to be made via neuromarketing of "brain-based" resources. Research has even shown that educators are very susceptible to neuromarketing.

In one study, teachers were more apt to believe the content of a resource if it included an image of a brain on it as opposed to just including the word brain in the title (that is why we included both on the cover of *Neuroteach*).[7] It is therefore imperative that teachers, who have little time to acquire resources that help translate "good research" into practice, remain optimistic of the promises of neuroscience from sources that have definitive footnotes and replicable research studies.

This book was written to aid in that process; to serve as a conduit between the research and the classroom teacher and school leader, who also recognize educational neuroscience as essential to enhancing student achievement. In doing this, we also think we have written a book that will be very valuable to all parents who are interested in seeing their children have an excellent education.

In all the discussions around PISA (Program for International Student Assessment), Common Core, and the standards students should be expected to meet to prepare them for the world they will inherit, what is missing is an equal emphasis on how students will learn the foundational knowledge and skills that will prepare them for their future. We should not just want students to hold information in their brain long enough to "pass" a standardized test. We should want core knowledge to remain with students and be foundational for the higher order, creative thinking that their future career choices will demand. An understanding of neuroscience, how to enhance student memory, attention, and motivation, and more importantly their own strengths and weaknesses as learners, does for students what Dr. Judy Willis suggests, "Lower the barriers but not the bar."[8]

In our work directing the Center for Transformative Teaching and Learning at St. Andrew's Episcopal School, and having the chance to present to and work with teachers, school leaders, researchers, parents, and policymakers in the United States, China, Dubai, and United Kingdom, we have learned a number of lessons when teachers and school leaders are introduced to the principles and strategies of educational neuroscience:

1. Teachers and school leaders recognize that the "science of learning" is not as complicated or inaccessible as they thought it would be;
2. Teachers and school leaders recognize the importance of differentiating their instruction because research shows this enhances student engagement and attention;
3. Teachers and school leaders develop an enhanced belief in or efficacy of themselves as professionals as they add the science of the brain to the other qualities that define the knowledge, skills, and mindsets expert teachers possess;
4. Teachers and school leaders quickly begin changing some of their practices to become more effective in their work with students.

The combination of these is a professionalization of practice, placing teachers at the cutting edge of an exciting, expanding field, unleashing waves of creativity, intellect, and collaboration. It is teacher manna.

As much as teachers are brain-changers for their students, our wish for the readers of *Neuroteach* is to begin to rewire each teacher's brain, to develop new neural pathways and connections informed by mind, brain, and education science. We recognize that this ambitious brain rewiring is no easy task because of two factors. First, teachers tend to teach how they were taught in school. Second, teachers tend to teach to their learning strengths.

In 1997, John Bruer's article "Education and the Brain: A Bridge Too Far" highlighted the unlikely union or linking of neuroscience and education.

Neuroteach continues to build the "bridge" between the lab and the class-room that Bruer described, but from a different direction, the everyday classroom-teacher experience. The "experts" on learning are those who spend each day in the classroom with students. Whether they know it or not, every teacher is a researcher. The number of observations and deci-sions they make about their class and their students is equal only to that of doctors, but they often do not keep current in educational advances at the same level we expect those in the medical profession to do. This is a gaping hole in the professionalization of teaching practice that needs to be filled. We therefore hope that this book, written by teachers who continue to be in the classroom and continue to work with students every day, leads to:

- Research in MBE science informing how teachers design their classes and work with each student: the often overlooked "just fine" kids and the academic "high fliers" too, not just the ones who are struggling.
- Research in MBE science becoming a required part of teacher train-ing and ongoing professional-development programs.
- Research in MBE science leading to an increased number of repli-cable school/university partnerships.
- Research in MBE science being used to validate and challenge peda-gogical practices.
- Research in MBE science providing a pathway for professional devel-opment that creates expert teachers.

Neuroteach is our effort to synthesize and translate research into next-day practices, and to make the case for increased collaboration and partner-ships between schools and universities that each share authority for better understanding how the brain learns and how learning happens. It is also a call to action. For all the great ideas about how best to teach, not enough are validated by mind, brain, and education science because not enough teachers and school leaders have an understanding of MBE principles and strategies informed by research. This is equivalent to a doctor not keeping current with the latest medical procedure. Would you consider having a heart procedure done by a surgeon trained in the 1980s who is still using the bulk of the methods he or she learned then? We now have the means to raise the professionalism bar for teachers and make this happen.

We love teaching and working with students. We also love the challenge of trying to figure out the body's must complex organ and its implications for learning. But when we were asked "for what purpose are you writing this book?" the answer we gave is that it is the same reason that we chose a career in education: to benefit each student who deserves to have a teacher, in every year of his or her academic journey, and at every point of his or her day, who believes in his or her potential, and who knows the science behind teaching, how brains learn, and how students thrive.

NOTES

1. "Leadership is second only to classroom instruction among all school-related factors that contribute to what students learn at school." Kenneth Leithwood, Karen Seashore Louis, Stephen Anderson, and Kyla Wahlstrom, "How Leadership Influences Student Learning," Learning from Leadership Project, The Wallace Foundation, 2004.

2. Abigail L. Larrison, "Mind, Brain and Education as a Framework for Curricular Reform" (PhD diss., University of California, San Diego; California State University, San Marcos, 2013).

3. Dr. Christina Hinton, Harvard University Graduate School of Education, personal communication, 2014.

4. R. JohnBull, Mariale Hardiman, Luke Rinne, "Professional Development Effects on Teacher Efficacy: Exploring How Knowledge of Neuro- and Cognitive Sciences Changes Beliefs and Practice" (paper presented at the American Educational Research Association conference, San Francisco, 2013).

5. Research Schools International led by researchers from Harvard's Graduate School of Education is one model for a university/school partnership.

6. Judy Willis, "How to Teach Students about the Brain," *Educational Leadership* 67, no. 4 (2009); Judy Willis, "What You Should Know about Your Brain," *Educational Leadership* 67, no. 4 (2009), retrieved from http://www.ascd.org/ASCD/pdf/journals/ed_lead/el200912_willis.pdf.

7. Annukka K. Lindell and Evan Kidd, "Consumers Favor 'Right Brain' Training: The Dangerous Lure of Neuromarketing," *Mind, Brain, and Education* 7, no. 1 (2013): 35–39.

8. Judy Willis, presentation to St. Andrew's Episcopal School faculty, 2013.

1

EDUCATIONAL NEUROSCIENCE
FOR ALL

Teaching without an awareness of how the brain learns is like designing a glove with no sense of what a hand looks like. If classrooms are to be places of learning, then the brain—the organ of learning—must be understood and accommodated.

—Leslie A. Hart, *Human Brain and Human Learning*

Early in our work with the science of learning, how the brain learns, we hit a cultural firewall. Anytime we found ourselves talking about the brain and learning, parents often construed that such science was only good for the struggling student, students who might be diagnosed with learning disabilities. However, what we quickly realized is how critical neuroscience is for all: the most advanced student, the often-overlooked "just fine student," *and* the struggling student. As an example, executive functioning, which is the ability to plan, organize, and execute, is critical for every person. It takes place in the frontal lobes of the brain, a region that is last to become highly developed, not until the mid-twenties at least. However, in the world of education, executive functioning is usually only connected to students as a "disorder."

All the way through schooling—through elementary school, through secondary school, through college, through a master's degree, and into a PhD—the prefrontal cortex is still developing. While there is a genetic component, this development, all the way through, will be affected by the environment, by experiences that student has, and by how the student reflects upon and unpacks those experiences. This is the concept of neuroplasticity, and it is something in which schools, for better or for worse, whether they sign up for it or not, play a role.

Look more closely at the executive functioning skills that schools can influence the development of: "problem-solving, prioritizing, thinking ahead, self-evaluation, long-term planning, calibration of risk and reward, and regulation of emotion."[1] These are skills that *all* students—the most advanced students, the often-overlooked "just fine students," *and* the struggling students—can benefit from being as good at as they possibly can be.

Figure 1.1. Educational neuroscience for all.

So executive functioning, a suite of skills crucial for learning, for jobs, and for life, a suite of skills from which all students can benefit at excelling, a suite of skills that schools can help grow by their deliberate actions, is relegated to something that schools dare not talk about for fear of, at best, the discussion being labeled as all about learning-disabled students, or, at worst, the school itself being labeled as an institution for learning-disabled students.

Schools have a window, in the case of executive function, at least a twenty-year window, where they can influence the rewiring of students' brains. The sad thing is that most schools either ignore or are ignorant of the research and just leave this neuroplastic brain development to chance. We tend to start this task pretty well in early elementary education, then we tend to drop the ball monumentally, more so the older students get, thinking they will just pick up this stuff as they go. So the real nugget of gold that is "educational neuroscience for all" is that we can help *all young people*, *at all ages of schooling*, the most advanced student, the "just fine student," *and* the struggling student; we can help them all to rewire their brains to become better learners and higher-achieving students. By deliberate actions we take as to how we teach, how we assess, how we guide learning, we can make this happen—for all students. The first step in jimmying this nugget free to see the light of day is being able to talk about it in public without the crucial words *"for all"* being summarily dismissed, and the horribly oversimplistic label "learning disabled" attached.

First, it is critical to differentiate between the terms "brain" and "mind." In our early experience, we used these terms interchangeably. However, when we finally had the chance to listen to Nobel Laureate Dr. Eric Kandel, well known for his work on memory, speak about his publication, *The Age of Insight*,[2] we recognized our inaccurate interchanging of the terms "mind" and "brain." So what did we do? We emailed Dr. Kandel, and he brought clarity to our thinking:

> This new science of mind is based on the principle that our mind and our brain are inseparable. The brain is a complex biological organ possessing immense computational capability: it constructs our sensory experience, regulates our thoughts and emotions, and controls our actions. It is responsible not only for relatively simple motor behaviors such as running and eating, but also for complex acts that we consider quintessentially human, such as thinking, speaking, and creating works of art. Looked at from this perspective, our mind is a set of operations carried out by our brain.[3]

As educators in this field, we sound a lot more believable when we say, "our mind is a set of operations carried out by our brain."

Second, we have found that discussions go much better when you talk about "excellent teaching" rather than "helping students learn best." One begets the other, but the former is something that every parent in the world wants—the best teaching that it is possible to get, regardless of school type, geographical location, or tax base. Whereas the latter tends to bring about a reactionary, "My child doesn't need any help with that; what are you implying?"

Third, there is a lot of research in the field of mind, brain, and education (MBE) science about what excellent teaching entails, and what it most definitely does not include as laid out in "the unconscionable list" and "top twelve" lists in chapter 3. The research is emerging, and the lists are not complete, but teachers, acting as teacher-researchers (which we discuss in chapter 12, "Teachers Are Researchers"), will play a crucial collaborative role in helping the lists evolve.

So we are at a point when we can start explaining educational neuroscience for all. We can help all students rewire their brains to increase how their minds can perform. We can do this through excellent teaching. And "excellent teaching" is not just a nebulous statement; it involves teachers doing *these* practices and not doing *those* practices. The exact nature of these practices will look different, varying with the contexts of each individual classroom, but there is now sufficient MBE research to create these lists.

And, yes, we do mean all students: all ages, all abilities. However high a flier a student is, educational neuroscience for all can make that student fly even higher (it can also help the student fly in interesting new directions, and, as an added benefit, get more sleep as he or she does so). The "just fine" student, the one who, well, is doing just fine, who "falls through the cracks" in most schools, the one whom education books typically are not written about, can increase his or her skills, knowledge, and confidence through educational neuroscience for all. For the struggling student, educational neuroscience for all can help increase his or her skills, knowledge, and confidence, too.

Educational neuroscience for all involves teachers doing *these* practices and not doing *those* practices. Educational neuroscience for all involves school and political leaders setting a bar that requires doing *these* practices and not doing *those* practices with students. Educational neuroscience for all involves parents demanding schools do *these* and not do *those* practices. The goals of this book are thus twofold: first, to convince you this is true; second, to show you what these practices are.

Assuming that we are now able to engage in a dialogue about how neuroscience really can benefit all students, the high flier, the "just fine," and the struggling student, we can start thinking about how this might be done. The first step is training teachers in a neurodevelopment model, which research says is a key step in improving student performance.

As of 2016, four neurodevelopmental training models exist: Neuroscience and the Classroom,[4] The Brain Targeted Teaching Model,[5] All Kinds of Minds,[6] and Brain and Learning/The NEA Foundation.[7] They have different levels of research, and different levels of practical application, but all have been important steps in showing that it is possible to get ideas from the world of academic research into a form where it is usable by teachers.

Furthermore, research by Professor Mariale Hardiman of the Johns Hopkins University School of Education, and creator of *The Brain Targeted Teaching Model*, suggests that doing so improves students' learning.[8] In *Neuroteach*, we want to build on this work to meet the needs of teachers who want to implement research-based strategies to transform their practice, assess the impact of what they try, and do so in a sustainable way.

One of the most important things these models give a teacher is a neurodevelopment lens through which to view each student, their course, or class on a thirty-thousand-foot overhead-view level, and the week-to-week or day-to-day implementing in the classroom level.

For example, let's juxtapose All Kinds of Minds[9] and Howard Gardner's theory of multiple intelligences. Research suggests that a common misinterpretation of Gardner's theory of multiple intelligences—that teachers should be tailoring instruction to meet each individual student's strengths (linguistic; logical-mathematical; bodily-kinesthetic; musical; interpersonal; intrapersonal)—is actually a neuromyth[10] (as is the idea that people are either left-brained or right-brained).[11] Teachers should not be tailoring instruction to meet each individual student's strengths. Instead, research suggests although each student has individual learning preferences, all students learn best when taught in a variety of modalities. The best modalities to use will vary from concept to concept. Teachers should differentiate based on content, not learning style.[12] This misrepresentation of Gardner's work is, unfortunately, quite pervasive.

A more correct interpretation of Gardner's theory of multiple intelligences is that individual differences exist—each person is better at some of them, worse at others. The All Kinds of Mind framework provides three levels of categorization to provide finer distinctions in a similar vein to Gardner's multiple intelligences, but this time in terms of neurodevelopment demands that might be placed on the brain; its eight broadest categories (called "constructs") are memory, attention, language, spatial ordering, temporal sequential ordering, neuromotor functions, social cognition, and higher-order cognition.

It was originally conceived as a way to characterize the strengths and weaknesses of struggling students with an eye to helping students identify their strengths and leverage them to address their weaknesses—a way to identify and address individual differences. This is good, and it definitely works as a tool to do this, but we found a much more powerful use for it that benefits all students.

Whenever we give All Kinds of Mind workshops, there is a real "aha!" moment—teachers realize that the greatest power of this framework is as a lens through which to see and manipulate their courses at the day-to-day, week-to-week, and thirty-thousand-foot level. Each academic discipline and each content area within a subject has neurodevelopmental demands that are germane to it. The All Kinds of Mind framework allows teachers to align the neurodevelopmental demands inherent in the material they want to teach with the neurodevelopmental demands of how they teach it and the neurodevelopmental demands of how they assess it. When they do this, it benefits all learners, not least because there is a "fairness" that students see and appreciate.

But it goes deeper than this. Viewing their classes through this lens tends to unleash innovative, creative teaching and assessing, which tends to foster student motivation and engagement, and it leads to teachers differentiating more, which helps all learners. It also helps teachers balance out the neurodevelopmental demands they are placing on their students day to day and week to week. For instance, this way, memory storage and retrieval and language processing do not get hit day after day after day.

Instead, *these* demands might be placed on a student for a while, before the teacher deliberately switches to placing *those* demands, all the while bearing in mind the demands inherent in the subject. Furthermore, by adding variety to the demands they are placing on their students, teachers are making their class more challenging while at the same time fostering engagement—more challenging because in order to get a top grade, students now have to master a greater variety of neurodevelopmental demands.

Struggling students may be asked to do things that they find themselves better at; the best and brightest might be made to struggle for a while; the "just fine" students will not be coasting along in a comfortable groove, but rather moving from tasks that they find easier to tasks that they find more challenging. Teaching teachers a neurodevelopmental framework both inspires and equips them to vary modalities of teaching and assessment, and differentiate based on content rather than learning style—all these, as we mentioned before, are factors that research says lead to increased learning.

By equipping teachers with a neurodevelopmental framework, a new lens through which to view their craft, we have made one step along the path of educational neuroscience for all. Two more steps come by using methodologies from the Top Twelve list and avoiding one on The Unconscionable List. More steps are outlined in the following chapters. But we want to end by taking one further look at how educational neuroscience for all applies to the best and the brightest, the highest of fliers, and thus to everyone.

Think of students who are good at listening to their teacher, reading their textbooks, and remembering what they hear and see. Typical school tests, that staple of grading, are pretty easy for them. How do you push these students? How do you build their resiliency? You could give them more to read, more to memorize, and maybe a shorter time in which to do it. But is this really stressing them? It might cause them stress, yes, but they fundamentally know that the task ahead of them is something that they are able to do. Maybe they are self-aware enough to know that it is even something they are good at.

The crunch of "not enough time!" to do something that you know you could do if you had a bit longer is different than the stress that comes from knowing that the task ahead is something that you are not good at, that failure, absolute failure, is a possible outcome. Don't worry; tolerable stress, as we learn in chapter 7, can be a good thing if it occurs in short duration and in a supportive community. How do we get the best and the brightest out of their comfort zone? How can we put them in situations where they realize that to succeed they will have to build new competencies, knowledge, and confidence? Knowledge of a neurodevelopmental framework gives a teacher a great toolkit to do this.

To answer this question fully, we need to explore another tendril of "educational neuroscience for all!"—the nature of intelligence. Humans have long attempted to define what intelligence is. Contemporary views see psychometric intelligence, the intelligence of IQ and standardized tests, as an important component of intelligence, but not the only component (this was the major theme of the 2014 Learning and the Brain conference in New York). A broader definition of intelligence certainly includes creativity[13] and maybe a category that we could call personality or social intelligence, which includes factors like resiliency, motivation, curiosity, and social cognitive ability.

Furthermore, intelligence does not reside in one spot in the brain, nor does creativity, but rather involves networks of parts of the brain, all working together. Scientists have now mapped out the brain network responsible for psychometric intelligence and also the brain network responsible for creative intelligence—and, unsurprisingly, they are different. The brain network for social intelligence is different still. This means that being good at one of these three forms of intelligence does not necessarily mean you are good at the others.

One of the huge ideas to come from the field of MBE science is that intelligence is not fixed at birth. The nature versus nurture debate has been settled; the answer is that it is a combination of both. Genetic differences and environmental effects, particularly in early childhood, lead to individual differences. We are all stronger or weaker in one form of intelligence than others. But neuroplasticity means that as we work with children in schools, we have the potential to help students rewire their brains to improve their performance, to some, but significant, degree, in all three of these intelligence areas.

Psychometric intelligence, creativity, and personality/social cognitive ability: which of these does traditional schooling emphasize and incentivize? We suggest that schools put too much of their emphasis on psychometric intelligence at the expense of the others, and it is time to redress that imbalance. If intelligence is a three-legged stool, it will always be able to sit there no matter how different the lengths of the legs are—but would you want to stand on it? So part of "educational neuroscience for all!" means developing a broader definition of intelligence in all students; for example, just because a student has high test scores doesn't mean he or she is creative. Neuroplasticity means we can address that, and a look at the work world students will one day enter means we should address that.

To emphasize why this is important, consider Tony Wagner's list of seven critical competencies, compiled following interviews with hundreds of business leaders to discover what skills young people need to be successful, to close what he calls "the global achievement gap."[14] These are the skills that are needed, but that are, Wagner amongst many others argues, too rare in those entering the workforce:

- Critical thinking and problem solving
- Collaboration across networks and leading by influence
- Agility and adaptability
- Initiative and entrepreneurship
- Effective oral and written communication
- Accessing and analyzing information
- Curiosity and imagination

Which of Wagner's "seven survival skills" rely exclusively on psychometric intelligence? Which don't? Where does creativity come into play? What about personality and social cognitive ability? Remember, there are different brain networks at work here, and having high psychometric intelligence does not necessarily mean you are strong in other forms of intelligence. Schools put a lot of focus on psychometric intelligence; but what about other forms of intelligence?

Think again of those students who are good at listening to their teacher, reading their textbooks, and remembering what they hear and see. Most assignments in most schools are pretty straightforward. School is challenging, but mostly because of the gargantuan quantity of work—particularly homework—that students have to do, as we discuss in chapter 10, not because of the cognitive complexity of individual tasks. They are getting good grades. But isn't their education selling them short? Think of all the other amazing, challenging, deeply engaging cognitive tasks we could be putting in front of them.

Imagine you were given the chance to design a school that really sought to develop students in the most critical knowledge, skills, and mindsets. What would it look like? What would you incentivize? What kinds of tasks would students do? Would it look like most schools today?

In chapter 10, "Homework, Sleep, and the Learning Brain," we introduce the toxic effect that Dr. Denise Pope at Stanford University calls "doing school." Educational neuroscience for all means changing our practice so that we make the fundamental shift from "doing school" to learning. Look at a kindergarten classroom. Intellectual curiosity and intrinsic motivation abound. The children here are not "doing school." There tends to be a palpable energy and passion around learning. Where does this go? How do we get it back at every grade level? What a fabulous challenge that would be.

Without looking back from this page, what are the *three* most salient points you take away from this chapter of *Neuroteach*?

What are *two* things you would like to do "tomorrow" with the information you learned from reading this chapter?

What is *one* question you have after reading this chapter?

NOTES

1. "Young Adult Development Project," retrieved from http://hrweb.mit.edu/worklife/youngadult/brain.html (accessed October 13, 2014).

2. Eric Kandel, *The Age of Insight: The Quest to Understand the Unconscious in Art, Mind, and Brain, from Vienna 1900 to the Present*, first edition (New York: Random House, 2012).

3. Eric Kandel, personal communication.

4. "Neuroscience and the Classroom," retrieved from https://www.learner.org/courses/neuroscience/.

5. "Mariale Hardiman's Brain-Targeted Teaching Model," retrieved from http://braintargetedteaching.org/.

6. "All Kinds of Minds," retrieved from http://www.allkindsofminds.org/.

7. "Online Courses // The NEA Foundation," retrieved from http://www.neafoundation.org/pages/courses/.

8. R. M. JohnBull, M. Hardiman, and L. Rinne, "Professional Development Effects on Teacher Efficacy: Exploring How Knowledge of Neuro- and Cognitive Sciences Changes Beliefs and Practice" (paper presented at the AERA conference, San Francisco, CA, 2013); M. Hardiman, R. JohnBull, L. Rinne, J. Pare-Blagoev, E. Gregory, and J. Yarmolinskaya, "How Knowledge from the Science of Learning Influences Teaching Practices and Attitudes" (in preparation).

9. Since 2007, 100 percent of St. Andrew's teachers have been trained in the All Kinds of Mind neurodevelopmental framework, which explains our choice here.

10. Paul Howard-Jones, *Introducing Neuroeducational Research: Neuroscience, Education, and the Brain from Contexts to Practice* (New York: Routledge, 2010).

11. Mariale Hardiman, *The Brain-Targeted Teaching Model for 21st-Century Schools* (Thousand Oaks, CA: Corwin, 2012).

12. Mariale Hardiman, presentation to CTTL Neuroeducation Leadership Institute, 2013.

13. Rex Jung, presentation at Learning and the Brain Conference, New York, 2014.

14. Tony Wagner, *The Global Achievement Gap: Why Even Our Best Schools Don't Teach the New Survival Skills Our Children Need—And What We Can Do About It*, first trade paper edition (New York: Basic Books, 2010).

2

A FORMATIVE ASSESSMENT

When I ask teachers to name the biggest obstacle to good teaching, the answer I most often hear is "my student" . . . Criticizing the client is the conventional defense in any embattled profession and these stereotypes conveniently relieve us of any responsibility for our students' problems—or their resolution.

—Parker Palmer, *The Courage to Teach*

Whether unconsciously or consciously, you bring to your reading of this book prior knowledge in mind, brain, and education (MBE) science. This knowledge could have emerged from your own educational experience, professional development, or reading selections. Prior knowledge is critical to learning. The brain likes to connect incoming information with information and experiences already stored in its long-term memory. However, it is also important that this knowledge is accurate. In the field of MBE, neuromyths abound and are unfortunately informing how teachers teach.

When beginning a new unit, or in this case a book, it is critical to find out what the audience already knows; this is something we often do with our students. Therefore, since this book is about translating research in how the brain learns to the instructional practice of educators, it is only fitting that before we take a deep dive into the research and strategies, that you take a nonthreatening assessment of what you already know about the learning brain.

One of the most underutilized yet critical teaching and learning strategies that MBE science validates is formative assessment. Providing students more frequent, nonthreatening, or low-stakes, feedback on their understanding is critical to memory consolidation. Yes, we are actually saying that we should provide students more frequent assessments. For example, as history and science teachers, we want to know what prior knowledge our students bring to our classes, and we also want to give them the best tool possible to help them store important information in their memory. As a result, we begin every unit with a formative assessment or pretest. It is therefore our hope that each reader takes the opportunity to gauge his or her current knowledge and prepare his or her brain for a deep-dive into the application of MBE science. By going to www.thecttl.org/neuroteach, you can complete the assessment shown below, as well as see the current aggregated scores of all readers.

It might seem counterintuitive to commence a book that is arguing for the next frontier for teacher training with a "traditional" true/false quiz. But it only looks like this; think about how we are using it. You are not being graded. The goal of this formative assessment is to help everyone learn and this quiz is priming your brain for what's coming in future chapters of *Neuroteach*. As this book will argue, in order to enhance a student's ability to secure into his or her long-term memory essential content determined by the Common Core, a school board, or classroom teacher, we must know what prior knowledge that student has. We are interested in learning what prior knowledge each of the readers of *Neuroteach* has before diving deeper into the book's content.

We also use formative assessments in the middle of a unit to give each of our students feedback on what they actually now know and what they do not know to help guide the studying they will do next, and also to give us feedback to adjust our teaching. Getting this feedback improves both learning and teaching. In this spirit, go to www.thecttl.org/neuroteach at any point you like to test and refresh your knowledge (as well as check how well you did if you filled out the paper version below). We will use your anonymous answers to help guide our future work.

Respond to each statement below with "T" for true or "F" for false:

_____ 1. The more a teacher knows about neuroscience, the more differentiated his or her instruction will be.

_____ 2. Students should be praised for their intelligence, not their effort.

_____ 3. The ability of the brain to change stops around the age of sixteen.

_____ 4. Human brains seek and often quickly detect novelty.

_____ 5. Humans use about 10 percent of their brains.

_____ 6. Integrating the arts into the curriculum enhances learning and understanding.

_____ 7. Informing students before an assessment that they will receive feedback/results sooner lessens their performance.

_____ 8. Providing students opportunities to self-correct wrong answers enhances retention of information.

_____ 9. Spaced instruction and studying enhance long-term memory consolidation better than mass instruction and review.

_____ 10. There are brain differences by race.

_____ 11. Listening to music with words while studying enhances a student's ability to learn material.

_____ 12. A student's emotions affect learning, memory, and recall of information.

_____ 13. Frequent, ungraded, formative assessments enhance memory consolidation.

_____ 14. Multitasking reduces memory consolidation.

_____ 15. Some students are left-brained and some students are right-brained.

_____ 16. Regularly changing the decorations and/or organization of a classroom enhances attention.

_____ 17. Individuals learn better when teachers teach and assess in their preferred learning styles.

_____ 18. In a class period, the information that is delivered first is what students remember best and the information that comes last is what the students remember second best.

_____ 19. Sleep enhances memory consolidation.

_____ 20. Learning is enhanced by challenge and inhibited by threat.

_____ 21. Having students memorize information is an outdated instructional strategy.

_____ 22. Providing students choice in their learning enhances engagement and deepens learning.

_____ 23. Brains are able to multitask.

_____ 24. The more teachers understand principles from educational neuroscience, the more they will believe in a student's ability to improve their academic performance.

We do not give pop quizzes that can impact the grade. Research says they do not help learning; research also says that formative assessments do. In John Hattie's book *Visible Learning*, which analyzes and ranks over eight hundred meta-analyses of studies relating to student achievement, "providing formative assessments" ranks the third-highest out of 815, and ranks the highest out of things in the domain of the teacher.[1] So we made the switch.

Part of the power of formative assessments is that they address "if you don't use it, you lose it" in a low-stakes way. Students need to use knowledge soon after encountering it to help them store it in their long-term memory. The act of trying to recall also helps memory storage. Spaced studying helps memory storage. Stress can impede memory storage and recall. All these factors, informed by MBE research, support our decision. As teachers, we found this to be a very simple and very effective switch to make. We have tried it in our classrooms, and it works. We have sought feedback from students about the change, and it works. We have spread the word to colleagues, sought their feedback, and it works.

What is more, this is good for all students, the high fliers, the "just fine kids," and the ones who are struggling, because it both helps students learn and helps them learn more in less time. Think, for example, of a student in the midst of law school or med school—wouldn't this be good for them? Knowing to create their own formative assessments as a study strategy to grow and guide their learning in a time-efficient way has to be good. And the confidence that comes from knowing they have this tool is powerful too.

If a teacher ever gives a pop quiz, he or she needs this book. If a teacher ever begins class by going over homework, he or she needs this book. If a teacher ever advises students to study for a test by primarily reading over their notes or their textbook, he or she needs this book. But this book is also written with school leaders and parents in mind—it is a guide for accountability. This is what research says is good teaching, and it is a research-informed do and do not list.

How does your child's teacher or your child's school measure up? Are they the educational equivalent of sending your child to a doctor who attempts to treat patients with leeches? The emerging new definition of teaching excellence, based on a convergence of ideas from neuroscience, cognitive science, psychology, and educational research, is all about the quality of the teacher, the quality of the school leadership, and the quality of the professional development—this book, this checklist, is how we begin, en masse, to demand it.

Without looking back from this page, what are the *three* most salient points you take away from this chapter of *Neuroteach*?

What are *two* things you would like to do "tomorrow" with the information you learned from reading this chapter?

What is *one* question you have after reading this chapter?

NOTE

1. John Hattie, *Visible Learning: A Synthesis of over 800 Meta-Analyses Relating to Achievement*, first edition (London and New York: Routledge, 2008).

3

THE TOP TWELVE RESEARCH-INFORMED STRATEGIES EVERY TEACHER SHOULD BE DOING WITH EVERY STUDENT

My point is that perceptual bias can affect nut jobs and scientists alike. If we hold too rigidly to what we think we know, we ignore or avoid evidence of anything that might change our mind.

—Martha Beck

Why wait until the end of this book to provide the reader with what he or she really wants to know, and that is: "What are the research-informed strategies that teachers should be using to enhance student achievement and the learning experience?" So in this chapter we will give you the top twelve. However, you need to read the remainder of the book to understand better the research and rationale for each. We will also give you what we call *the unconscionable list*. Knowing what we do about how students learn best, having research about what works and what doesn't work, there has to be a list of practices that it is now unconscionable that teachers still do.

THE UNCONSCIONABLE LIST (AKA THE DESPICABLE BAKER'S DOZEN): THINGS A TEACHER SHOULD NEVER DO AGAIN

1. Pop quizzes for a grade.
2. Starting a class by going over homework.

3. Ending a class by teaching all the way to the bell.
4. Coaching students to use passive studying techniques, such as reviewing for a test by just rereading their notes or textbook.
5. Defining kids by an individual style, such as "This person is an auditory learner, that person is a kinesthetic learner."
6. Varying the modality of teaching to match these perceived individual learning styles.
7. Applying simple labels to students, such as "lazy" or "smart," rather than making judgments based on observations.
8. Believing students have a fixed level of ability (despite their being in a time of great brain plasticity, able to work in ways that will rewire their brains to make them better learners and higher-achieving students).
9. Content delivery dominated by lecturing.
10. Assessment dominated by tests, particularly multiple-choice tests.
11. Always being the sage on the stage and never the guide on the side.
12. Praising achievement rather than effort.
13. Not recognizing the connections between emotion, identity, and health to learning.

While it is tempting to rush you forward to the top twelve list of strategies at this point, that would not be good practice. Instead, we would like you to use the space below to take a moment to reflect.

Take two minutes for this: What do you see? Don't try to interpret yet, don't ask questions, just say what you see in this list.

Take two minutes for this: What do you think? Don't pose questions yet, just say what you think based on what you have seen in this list.

Take two minutes for this: What do you wonder? Just say what questions this list raises for you.

THE TOP TWELVE RESEARCH-INFORMED STRATEGIES EVERY TEACHER SHOULD BE DOING WITH EVERY STUDENT

1. Class periods should be designed with an understanding that what students will recall most is what takes place in the first part of the class and what students will recall second most will take place in the closing minutes of class.[1]
2. Students should be given more frequent, formative, low-stakes assessments of learning.[2]
3. Students need more opportunities to reflect, think meta-cognitively, on their learning and performance.[3]
4. Students need to know that the pervasive way they choose to study is actually hurting their ability to learn for the long term and that self-testing is much more effective than reading one's notes.[4]
5. Students, parents, teachers, and school leaders need to understand that sleep is critical to memory consolidation. Without sufficient sleep we create a system that perpetuates the illusion of learning.[5]
6. Students need to know that "effort matters most," and that they have the ability to rewire their brain to make themselves better learners and higher-achieving students (the concept of "neuroplasticity").[6]
7. Students need more, but well judged, opportunities for choice in their learning, which enhances engagement and intrinsic motivation.[7]
8. Students need to love their limbic system and recognize the impact stress, fear, and fatigue have on the higher-order thinking and memory parts of their brain.[8]
9. Students need opportunities to transfer their knowledge through the visual and performing arts.[9]

10. Students need their teachers to vary the modality of teaching and assessment based on the content (as well as the time of day): What methods suit this topic best? What methods have I just used and will use soon so that I can provide a range of challenges? All students learn best when taught in a variety of modalities, and when the modality is chosen with the content in mind rather than the student.[10]

11. Students need to know the anatomy of their brain, especially the role the prefrontal cortex, amygdala, and hippocampus play in their learning.[11]

12. Students need frequent opportunities during the school day to play.[12]

We have learned from doing workshops with many groups of teachers that research from mind, brain, and education science upholds many things that teachers already consider good practice, but people tend to find the research behind it interesting—it all seems to make sense about *why* a particular method works. This is our "in." It makes it easier to digest and accept the cases where MBE science provides research that says a common practice from the standard canon of teaching is actually something we should not be doing. Fortunately, as you can probably see in this chapter, MBE science often suggests an alternative.

Think stakeholders. The top twelve list includes things the teacher should do and things the student should do. Teachers have an important and perhaps more active role in coaching students than they might be used to. Parents have an important role in supporting, coaching, and praising students in ways that at times will seem unfamiliar. School leaders and policy makers can decide to make excellent teaching, informed by research in the field of MBE science, a priority to improve learning and achievement for all students.

Take a look at the lists in this chapter again. We think the most striking point is that both the "stop doing this" and "do this" lists are dominated by *simple* changes. If the only changes we make in schools are the things listed in this chapter (MBE science goes much deeper, but even if these are the *only* changes we make), learning will be better for all students.

But it does not quite end there because teachers have an interesting role they can choose to play in this enterprise. The research base of MBE science is expanding. The lists above will, rightfully, evolve over time, and teachers can be a part of this. This could happen formally, through partnerships between research institutions and schools, maybe collaborating on studies of sufficient rigor to be published in peer-reviewed journals. But there is also a need for teachers to conduct less formal but just as necessary research. High-quality observational and qualitative data can be collected by all teachers. The results can be used to inform personal practice, school-wide practices, or future more in-depth research studies and collaborations.

The teacher has a role as researcher, but also in creating cohorts of interested people and collaborators in his or her school, expanding the circles of people whom the research touches, seeding the next studies. It is ground-up professional development. It is more than action research. We call it Action Leadership. We hope this book suggests possible avenues to explore, gives you an accessible beachhead into the literature base, and helps put you in touch with a network of like-minded colleagues.

Without looking back from this page, what are the *three* most salient points you take away from this chapter of *Neuroteach*?

What are *two* things you would like to do "tomorrow" with the information you learned from reading this chapter?

What is *one* question you have after reading this chapter?

NOTES

1. David A. Sousa, *How the Brain Learns* (Thousand Oaks, CA: Corwin, 2012); A. D. Castel, "Metacognition and Learning About Primacy and Recency Effects in Free Recall: The Utilization of Intrinsic and Extrinsic Cues When Making Judgments of Learning," *Memory and Cognition* 36, no. 2 (2008): 429–37.

2. Henry L. Roediger III and Andrew C. Butler, "The Critical Role of Retrieval Practice in Long-Term Retention," *Trends in Cognitive Sciences* 15, no. 1 (2011): 20–27, retrieved from *ScienceDirect* (accessed March 21, 2014); Jeffrey D. Karpicke and Janell R. Blunt, "Retrieval Practice Produces More Learning Than Elaborative Studying with Concept Mapping," *Science* 331, no. 6018 (2011): 772–75, retrieved from www.sciencemag.org (accessed March 21, 2014).

3. Donna Wilson and Marcus Conyers, *Five Big Ideas for Effective Teaching: Connecting Mind, Brain, and Education Research to Classroom Practice* (New York: Teachers College Press, 2013).

4. Jeffrey D. Karpicke and Henry L. Roediger, "The Critical Importance of Retrieval for Learning," *Science* 319, no. 5865 (2008): 966–68; Nate Kornell and Lisa K. Son, "Learners' Choices and Beliefs about Self-Testing," *Memory* 17, no. 5 (2009): 493–501; Peter C. Brown, Henry L. Roediger, and Mark A. McDaniel, *Make It Stick: The Science of Successful Learning* (Cambridge, MA: Belknap Press, 2014).

5. Dr. Robert Stickgold, "Sleep Memory and Dreams: Fitting the Pieces Together," TEDxRiverCity, June 8, 2010, retrieved from http://www.youtube.com/watch?v=WmRGNunPj3c; Ullrich Wagner, Steffen Gais, Hilde Haider, Rolf Verleger, and Jan Born, "Sleep Inspires Insight," *Nature* 427, no. 6972 (2004): 352–55, retrieved from *NCBI PubMed*; Mary A. Carskadon, "Sleep's Effects on Cognition and Learning in Adolescence," *Progress in Brain Research* 190 (2011): 137–43, retrieved from *NCBI PubMed*.

6. C. Hinton, K. Fischer, and C. Glennon, "Mind, Brain, and Education: The Student at the Center Series," *Mind, Brain, and Education*, March 2012, retrieved from http://www.studentsatthecenter.org/sites/scl.dl-dev.com/files/Mind%20Brain%20Education.pdf; Carol S. Dweck, *Mindset: The New Psychology of Success* (New York: Random House, 2006); Carol S. Dweck, *Self-Theories: Their Role in Motivation, Personality, and Development (Essays in Social Psychology)* (Philadelphia: Psychology Press, 1999); J. A. Mangels, B. Butterfield, J. Lamb, C. Good, and C. S. Dweck, "Why Do Beliefs About Intelligence Influence Learning Success? A Social Cognitive Neuroscience Model," *Social Cognitive and Affective Neuroscience* 1, no. 2 (2006): 75–86; Marina Krakovsky, "The Effort Effect," *Stanford Magazine*, March/April 2007.

7. Mariale Hardiman, *The Brain-Targeted Teaching Model for 21st-Century Schools* (Thousand Oaks, CA: Corwin, 2012); Daniel Willingham, *Why Don't Students Like School? A Cognitive Scientist Answers Questions about How the Mind Works and What It Means for the Classroom* (San Francisco: Jossey-Bass, 2009); P. M. Miller, "Theories of Adolescent Development," in J. Worell and F. Danner, eds., *The Adolescent as Decision-Maker* (San Diego, CA: Academic Press, 1989); Mariale Hardiman and Glenn Whitman, "Assessment and the Learning Brain: What the Research Tells Us," *Independent School Magazine*, Winter 2014.

8. Lars Schwabe and Oliver T. Wolf, "Learning under Stress Impairs Memory Formation," *Neurobiology of Learning and Memory* 93 (2010): 183–88; Barbara L. Fred-

rickson and Christine Branigan, "Positive Emotions Broaden the Scope of Attention and Thought-Action Repertoires," *Cognition and Emotion* 19, no. 3 (2005): 313–32; "Neuroscience and the Classroom: Making Connections," funded by Annenberg Learner, produced by Harvard-Smithsonian Center for Astrophysics Science Media Group, 2011, retrieved from www.learner.org/courses/neuroscience; T. Grindal, C. Hinton, and J. Shonkoff, "The Science of Early Childhood Development: Lessons for Teachers and Caregivers," in B. Falk, ed., *In Defense of Childhood* (New York: Teachers College Press, 2011); J. P. Shonkoff, W. T. Boyce, and B. S. McEwen, "Neuroscience, Molecular Biology, and the Childhood Roots of Health Disparities: Building a New Framework for Health Promotion and Disease Prevention," *JAMA* 301 (2009): 2252–59.

9. Luke Rinne, Emma Gregory, Julia Yarmolinskaya, Mariale Hardiman, "Why Arts Integration Improves Long-Term Retention of Content," *Mind, Brain, and Education* 5, no. 2 (2011): 89–96, retrieved from *Wiley Online Library* (accessed March 21, 2014); Hardiman, *The Brain-Targeted Teaching Model for 21st-Century Schools.*

10. Paul Howard-Jones, *Introducing Neuroeducational Research: Neuroscience, Education, and the Brain from Contexts to Practice* (New York: Routledge, 2010); Hardiman, *The Brain-Targeted Teaching Model for 21st-Century Schools*; "Neuroscience and the Classroom: Making Connections"; Karpicke and Blunt, "Retrieval Practice Produces More Learning Than Elaborative Studying with Concept Mapping"; Hardiman and Whitman, "Assessment and the Learning Brain."

11. Judy A. Willis, "How to Teach Students about the Brain," *Educational Leadership* 67, no. 4 (2009); Judy A. Willis, "What You Should Know about Your Brain," *Educational Leadership* 67, no. 4 (2009), retrieved from http://www.ascd.org/ASCD/pdf/journals/ed_lead/el200912_willis.pdf.

12. Stuart Brown, *Play: How It Shapes the Brain, Opens the Imagination, and Invigorates the Soul* (New York: Avery Trade, 2010); Willingham, *Why Don't Students Like School?*; L. S. Vygotsky, *Mind in Society: The Development of Higher Mental Processes* (Cambridge, MA: Harvard University Press, 1978).

4

HOW MUCH DO WE NEED TO KNOW ABOUT THE BRAIN?

We are accustomed to see men deride what they do not understand, and snarl at the good and beautiful because it lies beyond their sympathies.

—Johann Wolfgang von Goethe

Research suggests that knowledge of educational neuroscience is a powerful commodity that leads to higher student achievement, engagement, and motivation. But what should people know? And who would benefit from knowing it? We believe that the key is not simply knowing about the biology or architecture of the brain, but rather knowing about brain architecture in the context of its direct relationship to learning. And who should know this? Students, teachers, parents, school leaders, and policy makers—all the stakeholders in education. The need-to-know nuggets are few in number and profoundly linked to actions that will improve education for all.

I. Brain plasticity and the rewiring process as it relates to these structures:

- Neurons
- Neural pathways
- Dendrites
- Axons
- Synapses

Figure 4.1. What should teachers know about the brain, the organ of learning that each of their students brings to class every day?

Neurons are the basic information-processing unit, and look a bit like trees. They receive signals from the branches (dendrites) and, if the signal is above a certain threshold, send an electrical signal along through the trunk (the axon). At the end of the axon is the synapse—a gap between neurons. Chemicals carry the signal across the synapse to the dendrite branches of other neurons. When our brains are firing—which they always are—electrical signals travel from neuron to neuron to neuron, branching out to different areas of the brain along these neural pathways.

The branchiness of the dendrites means that each neuron is connected to many other neurons. Thus the brain is made up of networks of interconnected neurons. The exact formation and arrangement of this interconnected, three-dimensional tapestry is constantly changing throughout our entire lives, though not always at the same speed. Changes are particularly rapid during the years in which students are at school. While there is a significant genetic component to the architecture of each individual's brain, the changes that happen are influenced by the environments and experiences to which we expose our brains. By deliberate effort—working harder and smarter—we can influence the type of changes that take place. This is the concept of neuroplasticity.

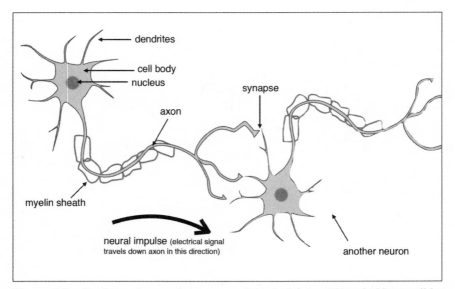

Figure 4.2. Basic anatomy of a neuron. Adapted from "Simple Neuron" by Fabuio, openclipart.org.

Teaching practices, assessment practices, study habits, the design of schools, the structure of the school day, the social and emotional environment of the school—all these factors influence how students' brains get rewired, and do so whether we seek to deliberately manage them or not. Perhaps the most fundamental premise of mind, brain, and education (MBE) science is that we can, and must, manage these factors in ways that promote the kind of changes that lead to stronger, happier, more motivated, more resilient, higher-achieving students.

II. Myelination[1]—Deliberate practice leads to increased speed of recall

The long axon trunks of the neurons are covered in fatty sheaths of myelin. Deliberate practice leads to these sheaths becoming thicker, which increases the speed that signals travel along them. The more you practice, the more a neural pathway fires, the more myelin builds up; this process is called myelination. The upshot is that working hard and working smart makes your brain faster, and increases your speed of recall.

III. Parts of the brain and their relationship to learning

The brain is a complex organ, and all cognitive tasks involve multiple parts working together, as shown in the visually stunning images from modern brain imaging techniques such as fMRI (functional magnetic resonance imaging) that, even if we do not understand the details, have become familiar in the public eye.

One powerful illustration of this is a short video available online that shows the brain's response, in slow motion, to hearing just a single word. There is a sequence of flashes of color—lighting up spots *here* and *here*, on both sides of the brain, then fading again as brain regions over *there* light up, then *there* . . . This simple sounding task, hearing just one word, involves a complex, highly coordinated, very nuanced cognitive response. It also explodes the myth that we are right-brained or left-brained people—all tasks use brain networks that involve both brain hemispheres—as well as the myth that we only use 10 percent of our brain.

Our other favorite brain image shows the response in different people's brains to the simple task of tapping one finger. There are obvious significant differences. It suggests that not only is the brain's response to any task very complex, it is also personalized—while we recognize general patterns of brain networks involved in certain tasks that hold true for the majority of people, each individual brain works differently than any other brain. My brain is different from yours, a function of both genetics—the brain we were born with—and environment—how it has grown, and been pared and rewired in response to all the experiences that it has had.

Our brains are complex and varied. What follows, then, is a list that is far, far from exhaustive, and which is a gross simplification—an oversimplification perhaps. So why include it? We feel this gentle primer might help us begin to understand the tip of the mammoth majestic iceberg of the brain's role in learning—and maybe whet your appetite for more.

(a) The Amygdala—As It Relates to Stress and Learning

The amygdala is a key part of the limbic system involved, among other things, in the brain's perception and processing of emotion. In our simplified model, to help us look at the brain's response to stress, it plays a role as the emotional switching station: too much stress impedes learning because the amygdala sends incoming information from your senses to the primordially hardwired reactive "fight, flight, freeze" part of the brain, rather than the reflective part of the brain, the prefrontal cortex, the function of which is explained below.

Getting sensory information to your prefrontal cortex is crucial for learning, and the key is the role of the amygdala and how you cope with stressors. Working at this leads to myelination of the neural pathway from the amygdala to the prefrontal cortex, meaning that doing so becomes easier over time. Research shows that moderate levels of stress in short, defined periods and in a supportive environment help rewire the brain to deal more effectively with stress in the future. The right amount of stress in the right conditions does indeed make you stronger.

(b) Hippocampus—As It Relates to Memory

The hippocampus plays a role in memory and spatial learning. It is a gateway "through which all information must pass before it can be memorized."[2] Research suggests that new neurons are created in the hippocampus throughout a person's entire life,[3] a process called neurogenesis. This is one form of neuroplasticity. Three other key ones are strengthening, pruning, and remodeling of neural networks during learning.[4] As a student learns, some neurons are triggered and others are not. The pathways of those that are used get strengthened, other neural connections are formed, and pathways that are not used get weakened or pruned away—"use it or lose it."

The basis of memory formation lies in these four interlinked neuroplasticity mechanisms: creating neurons, strengthening used pathways, creating new pathways, and pruning away unused pathways. The act of learning, therefore, plays a huge role in shaping how our brain forms enduring memory, and indeed in shaping the architecture of our brain itself. We should therefore be very purposeful about how we go about the act of learning.

(c) Prefrontal Cortex—As It Relates to Higher-Order Cognition and Executive Functioning

The prefrontal cortex plays a key role in many brain tasks, including executive functioning tasks: evaluating information, forming a plan, executing that plan, analyzing how that plan is going and making adjustments, and gauging a satisfactory endpoint. It also plays a critical role in higher-order thinking tasks. Neuroplasticity in the prefrontal cortex continues into a person's mid-twenties—so working at being better at planning and executing tasks (such as using scaffoldings and then removing those scaffoldings bit by bit) will, for a long time, help you become better at planning and executing tasks. Research also suggests that working to strengthen the prefrontal cortex also may help deal with stress.

IV. Neuromythbusting

While acquiring knowledge of the brain, one must deal with some annoyingly lingering neuromyths that need to be busted. Foremost are the following:

Brain development has finished by the time children reach secondary school.

FALSE—Your brain makes new neurons and new neural pathways throughout your life, and significant neuroplasticity continues in some areas of your brain beyond even college years.

Cognitive ability is solely due to your genes.

FALSE—While there is a significant genetic component to cognitive ability, the environment you are in and the experiences you have make a difference to how you brain grows and changes.

You cannot control how your brain grows.

FALSE—The types of work you demand of your brain help shape the creation, strengthening, and pruning of neural pathways in different regions of your brain. Therefore, the quality of effort and amount of effort you put in help rewire your brain.

We only use 10 percent of our brain.

FALSE—We use the vast majority of our brain, with many different regions working together to accomplish the tasks it is set.

People tend to be either left-brain people or right-brain people—this hemispheric dominance being responsible for their individual differences as learners.

FALSE—We all use both hemispheres of our brain. Most tasks we set our brains use different parts that are often located on different sides of the brain, so that most tasks use both sides of the brain.

Individuals learn better when they receive information in their preferred learning style (e.g., auditory, visual, kinesthetic).

FALSE—It is true that individual learners show preferences for the mode in which they receive information (e.g., visual, auditory, kinesthetic), but there is no evidence, despite many studies, to show that they learn better when they receive information in their preferred learning style. There is, however, evidence to show that different content topics are best learned using particular modes of teaching (e.g., visual, auditory, kinesthetic). Thus it should be the curriculum that is the driving force for selected modes of teaching, not trying to meet the perceived needs of individual learners in the classroom.

Emotions are a proven distraction from learning and should be kept apart from cognitive thinking.

FALSE—Cognitive and emotional areas are integrated in the brain, so that emotion, whether we like it, acknowledge it or not, is necessarily integrated with learning.

When we sleep, the brain shuts down.

FALSE—The brain is still active when we sleep, and certain crucial brain tasks, including ones associated with memory storage, only happen during this time. Sleep is vital for learning.

Our brains are able to multitask.

FALSE—Our brains cannot multitask. Instead, they flip back and forth, working on one task, then the other. However, there is a transaction cost for doing so, which means our brains work less efficiently. Studies have shown that people who self-identify as being able to multitask are no better at doing so than those who say they cannot.

Students should be praised for their achievement rather than their effort.

FALSE—It is the other way around. Associating success with their effort, not smartness, is key to helping students learn to persist. If a student believes intelligence is mostly a matter of effort, they are more likely to be motivated to exert effort, attempt difficult academic tasks, and persist despite setbacks, confusion, and failure. Making achievement the focus of praise tends to create a mindset of giving in when difficulties arise, and the path of successful achievement is blocked.

The more teachers know about the architecture of the brain and its direct relationship to learning, the more confident they will be in using MBE research to inform their practice. This is more than what we believe, this is also supported by research.[5] But we would dare to be bolder and suggest the following:

The more school leaders and policy makers know, the more helpful they will be in supporting this change. The more parents know, the better they will understand the frustrations and thrills they witness along the epic journeys of their children's learning, and the better able they will be to support them. The more students know, the more empowered they will be.

Belief in brain plasticity is critical not just for teachers and students, but for all stakeholders in education. Knowing how the biology and architecture of the brain can be changed through deliberate practice and strategies has the power to change study practice, teaching practice, school leadership, and educational policy; it has the power to shift the needle toward better education for all.

Without looking back from this page, what are the *three* most salient points you take away from this chapter of *Neuroteach*?

What are *two* things you would like to do "tomorrow" with the information you learned from reading this chapter?

What is *one* question you have after reading this chapter?

NOTES

1. See Daniel Coyle, *The Talent Code: Greatness Isn't Born. It's Grown. Here's How* (New York: Bantam Books, 2009), 30–35.

2. E. Maguire, R. Frackowiak, and C. Firth, "Recalling Routes Around London: Activation of the Right Hippocampus in Taxi Drivers," *Journal of Neuroscience* 17 (1997): 7103–10.

3. Elodie Bruel-Jungerman, Claire Rampon, and Serge Laroche, "Adult Hippocampal Neurogenesis, Synaptic Plasticity and Memory: Facts and Hypotheses," *Reviews in the Neurosciences* 18, no. 2 (2007): 93–114.

4. Elodie Bruel-Jungerman, Sabrina Davis, and Serge Laroche, "Brain Plasticity Mechanisms and Memory: A Party of Four," *The Neuroscientist: A Review Journal Bringing Neurobiology, Neurology and Psychiatry* 13, no. 5 (2007): 492–505.

5. R. M. JohnBull, M. Hardiman, and L. Rinne, "Professional Development Effects on Teacher Efficacy: Exploring How Knowledge of Neuro- and Cognitive Sciences Changes Beliefs and Practice" (paper presented at the AERA conference, San Francisco, 2013).

5

A MINDSET FOR THE FUTURE OF TEACHING AND LEARNING

Any [person] could, if [they] were so inclined, be the sculptor of [their] own brain.

—Santiago Roman y Cajal

One word. Three letters. One of the most important words in education. It is my favorite word to say to doubting students as well as skeptical teachers. Do you have a word in mind?

If you have gotten to this point in this book, you have already seen the word eight times and we are going to try to make the case that this one word is arguably the most impactful word for every teacher, school leader, and parent who wants to help each student meet his or her peak potential as a learner and as an individual. It is also a word, or mindset, supported by research. But it needs to be used with deliberate care.

Glenn first came across this word as an aspiring NHL ice hockey player when his mother responded to his saying, "I *can't* improve my slapshot" by using this word and then sending him outside to practice more. She had a similar response when he once proclaimed, "I *can't* juggle a soccer ball a hundred times." In each case, Glenn's "can't" was answered by his mom's "*yet*."

Now what we hope this chapter and book do is to begin a movement, or better still, to create a "yet sensibility" among teachers, students, and school leaders and for it to overflow into each of the various worlds we all inhabit in schools.

Figure 5.1. The power of "yet." Frazz © 2015 Jef Mallett. Distributed by Universal Uclick. Reprinted with permission. All rights reserved.

This word became central to our interactions and instructional practice with students as well as colleagues after reading Stanford University professor Carol Dweck's *Mindset: The New Psychology of Success*.[1] There are few researchers in behavioral psychology, one of the arms of the transdisciplinary field of mind, brain, and education (MBE) science, whose work has been more impactful to educators than Dweck's. It is hard to go to an educational conference, especially those that deal with the brain and learning, that does not reference her work. Dweck's research suggests mindset is often classified as one of the most critical "noncognitive" skills that students must cultivate to meet their potential as learners and as individuals.

But, before going any further, we flatly reject categorizing an individual's mindset as "noncognitive" or "soft skills"—terms that undermine the importance of the very things they try to highlight. Clearly they are cognitive skills that undoubtedly involve many brain systems working together, some combination of genetics and learned ability as the growing brain interacts with its environment. This means, as teachers, we can work toward growing these skills. And we should, because "soft" does not do justice to how greatly these skills can affect student performance. In fact, research continues to show the correlation between a student's mindset and his or her academic performance, and the ability to think and work at the highest level and in the face of the most difficult learning challenges.

A "yet sensibility" aligns well with the research on grit and resilience of University of Pennsylvania psychology professor Angela Duckworth, whose excellent work has significant implications for what "great" education might mean, and has often been connected to Anders Ericsson's "ten-thousand-hour rule" that was popularized by Malcolm Gladwell in his book *Outliers*. It is also linked closely with the field of mindfulness training that is beginning to be implemented in a growing number of schools.

But merely enhancing one's effort does not ensure a growth mindset. Nor does simply shifting from praising achievement to praising effort—though this is important, but more nuanced than commonly thought. As Dweck points out, having a growth mindset also requires the development of clearly defined strategies for improvement and the enlistment of support, advice, and guidance from others. The thoughtfully selected, reflectively evaluated, and iteratively tweaked use of strategies has been a common theme in this book, both for teachers and students, as has been the idea of working in collaboration rather than isolation.

The vital nuance of praising effort is that blindly praising effort in a "well done for working hard at this!" kind of way is not what is needed—praise should be linked to strategies, and strategies that have worked. If they have not worked, another kind of discussion is needed—one that acknowledges the effort made so far but moves on to think about what can be done next. It is more than working hard, it is more than working hard even using strategies, it is working hard using strategies in a smart way.

Moreover, important to having a growth mindset is the acknowledgment that we as individuals often switch between a fixed and a growth mindset depending on the situation. Rarely is someone solely one or the other.[2] As your day winds on and the environment, situations, or demands change, so your type of mindset is likely to change too. And all this will vary from person to person. But, thankfully, we can work to affect what mindset we gravitate toward in what situations—there is, unsurprisingly, brain plasticity at work here, too. The key to doing so is using strategies in a reflective manner, ideally with mentorship.

These factors are a reminder of the critical partnership and relationship that needs to exist between teachers and their students and the shared authority each brings to the learning process. They are also a reminder that we are all, brain-wise, works in progress, and that the amount and quality of effort we put in really does make a difference.

CULTIVATING A "YET SENSIBILITY"

The challenge for educators is how to cultivate a growth mindset, or "yet sensibility," within each student. First, it begins with the teacher who must have high expectations for all his or her students, and who must inherently believe that every single one of them can improve, which means forgoing labeling students.[3] Labels like "lazy" or "bright" mask a more nuanced perspective that recognizes that each individual student has a complex and ever-shifting profile of strengths and weaknesses.

So we must constantly evaluate and reevaluate each student's learning preferences through direct observations—allowing for preferences to change over time as each student's brain develops in environments that include our teaching. This even holds true for popular labeling around learning styles, where labels like "kinesthetic learner" often become a self-fulfilling prophecy; yes, the student may have strengths in this area, but let's take that more nuanced view that includes his or her other strengths and limitations and allows room for neuroplastic change.

Belief in plasticity and adaptability, knowing that you can learn, apply, evaluate, and tweak new strategies to address new situations, a true growth mindset, may be pivotal when we think of the unfolding lives of our students. Consider this. It has been suggested that today's students will have up to seven different career changes in their lifetimes.[4] Whether or not this number is accurate, it is certain that many of the careers today's precollegiate students will occupy have not been invented yet. For each of these jobs, they will need to learn new skills that they do not yet know. If students understand themselves as learners as well as their current yet evolving strengths and weaknesses, and have a growth mindset, they will be able to learn what is new to them as well as apply what is old to new situations.

"Yet" is not just a word for the future. As a history teacher, Glenn imagines many historical moments when "yet" changed the course of the past. You might picture in your mind and almost hear President Lincoln respond to one of his cabinet members in 1863 who declared, "You can't free the slaves" with President Lincoln declaring, "yet." Glenn was reminded of the "yet sensibility" as he read through one of his student's oral history projects in which she interviewed a Rosie the Riveter from World War II. At some point, the interviewee must have responded to an individual who declared, "A woman cannot do the work of a man who is off fighting in Europe or Japan" with the resounding word, "yet."

But getting to a "yet sensibility" is not easy. It is much quicker for a student to declare he or she "can't" do something and then give up or hope the teacher will provide the answer. But we know that struggle, challenge, and deliberate difficulty are really good for learning, especially for getting something to stick in one's memory. Challenge in school is important—without it, boredom reigns supreme. We just need to carefully scaffold it, maintaining the challenge but providing accessibility to it, "lowering the barriers, not the bar" as Dr. Judy Willis told us. Having a growth mindset is important for successfully meeting the challenge. fMRI scans show increasingly more brain activity when a student maintains a growth rather than a fixed mindset in the face of a learning challenge.

This is where increasing teachers' and students' knowledge about how the brain works and learns (as presented in chapter 4) becomes increasingly more important. Student-efficacy studies have shown that when students are taught how deliberate practice changes the neurology of their brains, their level of achievement increases and they have a greater belief in their capacity as learners.[5] The research that is central to truly creating a "yet" sensibility among students centers on brain plasticity: the idea that the brain is changeable; that through deliberate effort new neural pathways can be created; and that also through deliberate practice the process of myelination will thicken and speed up the neural pathways used, allowing students more easily to recall factual and procedural knowledge.

We have found that merely sharing the theory around growth mindset and brain plasticity is not enough. This can certainly be done with some success by having students read articles such as "You Can Grow Your Mindset" or by watching TED Talks from Carol Dweck and Angela Duckworth.[6] In addition, we have found that showing students how the brain is wired and can be changed through what Professor Mark McDaniel calls "effortful learning"[7] enhances their belief in their ability to change their brain.

But explicitly teaching students the architecture of the brain is taking the instruction to a level that is often not attempted in most schools. This is an empowering moment in a student's learning journey and should not be the sole responsibility of the biology teacher. Therefore, we should teach students about cell bodies, axons, dendrites, synapses, and myelination, as we discussed in chapter 4, and their connection to both old and new learning. Many students find this a fascinating insight into this vital organ they use so much every day, and which puzzles, frustrates, and amazes them with varied measure.

A barrier to developing a "yet sensibility" has been the belief that the science of the brain is too daunting to learn and inaccessible to those not trained in fields such as neuroscience. However, there is a growing collection of well-researched resources, such as *Neurocomic* by Hana Ros and *The Owner's Manual for Driving the Adolescent Brain* by JoAnn Deak, that now make the science of the brain much more accessible and playful (what we see as an underappreciated educational methodology) to students and teachers.

Moreover, university researchers such as Dr. Kurt Fischer and Dr. Christina Hinton at Harvard University have developed free, online courses through the National Education Association's Foundation titled "Brain and

Learning"[8] and the Annenberg Foundation course, "Neuroscience and the Classroom."[9] These are both research-backed resources that teachers and school leaders will find extremely useful.

Getting to "yet" begins after what we hope for each of our students is a good night's sleep (eight to nine hours), as they roll out of bed to face the opportunities and challenges of each new day. It is harder to do "yet" when you feel like a zombie, or feel as though you are just "doing school,"[10] so teachers also have some role to play here, by making sure they design and assign homework excellently, as we discuss in chapter 10. Getting to "yet" is also made immeasurably harder by identity threat—another reason for diversity and multicultural work, in the broadest sense these terms are used, encompassing all identifiers, to be a key part of education and teacher professional development.

IT TAKES PRACTICE

Perhaps the most important factor in building a "growth mindset" or "yet sensibility," however, is stunningly simple to say. It takes practice. The research, from Carol Dweck and others, is compelling: if we want students to achieve more, if we want to improve the quality of our teaching, if we want to improve our abilities as school leaders, it takes practice. Lots of practice. And not just any type of practice.

Over the years, scientists have explored the question of what made great individuals like Mozart. What research shows is that individuals who are the elite in their careers get there because of deep and deliberate practice that is defined by Daniel Coyle in *The Talent Code* as: "Working on technique, seeking critical feedback, and focusing ruthlessly on shoring up weaknesses."[11] Later we will discuss what deliberate practice might mean in the context of teaching and learning.

Malcolm Gladwell popularized "the ten-thousand-hour" rule, saying that it takes roughly this amount of practice to achieve mastery in a field.[12] While Gladwell's work has been a lightning rod, drawing some criticism, we can draw some interesting conclusions from it. First, given the nature of what is being measured and natural differences among people, measurable accomplishments will vary significantly from person to person and context to context. For example, Anders Ericsson has said that, "in the world of classical music it seems that the winners of international competitions are those who have put in something like 25,000 hours of dedicated, solitary practice—that's three hours of practice every day for more than 20 years."[13]

However, ten thousand hours may perhaps be taken as a reasonable average, and a sign that significant practice is needed.

Second, as Gladwell did his statistical analysis he found that a cohort of "naturally gifted" individuals did not emerge. We might imagine that there would be a number of people who rose to the level of mastery without putting in as much practice as the rest, but Gladwell did not find this to be the case. There is a direct relationship between hours of practice and achievement—the idea of "naturally talented" people who can take a shortcut to mastery without putting in the hours of practice is a myth not backed up by evidence. There are no shortcuts.

But while there are no shortcuts, there might well be "longcuts" that students may stray into. If we glance back at Coyle's definition of deliberate practice, we can imagine how one could be more efficient or less efficient at the important ongoing, iterative "seeking feedback and shoring up weaknesses" loop—and we can all picture in our minds students at each end of this efficiency spectrum. Having each student recognize what high-quality deliberate practice means for them in the context of his or her own individual learning strengths and weaknesses, and the context of the academic tasks they are being set, is a crucial part of them being effective, efficient learners.

It behooves us as teachers to help our students learn this skill. Given the significant plasticity of the brain throughout all school ages, this is good teaching no matter what the grade level. This is not about lowering standards or making things soft, less challenging. This is actually about setting a high, maybe higher, bar, but giving students scaffolding to help them reach it—scaffolding that can be peeled away as competency grows.

Does this talk of an extended period of deliberate effort sound hard? There is also the matter of what Professor Duckworth calls "grit"—the relentless determination in the face of the vast amount of practice ahead and the inevitable obstacles that will arise to keep going.[14]

While we do not expect our students to put ten thousand hours into studying for each of their assessments, the total number of hours they will spend on their schooling—in class plus at home—is much, much greater than this. In each student's own personal journey toward being "elite" historians or scientists, each of them has the ability to aid his or her journey, to make the best use of their hours, by preparing smarter with deliberate practice that works for them.

Research from MBE science can inform the strategies students use—increasing their effectiveness and efficiency. For example, using self-testing (rather than rereading the textbook) while avoiding switching between tasks to check what is happening on Facebook, Twitter, Snapchat, or the next best piece of social media is an important research-informed strategy students should be using. Deeply embedding the idea of "strategies" in each student's brain is what makes a difference. Students must buy into the idea of using strategies each day as they learn; that strategies are something that they have control over; that they have access to people around them who will advise them on choosing, using, evaluating, and tweaking strategies; that strategies will evolve over time as they themselves grow so they should not be seen as crutches they will necessarily carry forever; and that strategies really do improve learning. Strategies speak to something more than just working hard.

"I will work harder" is something heard quite often by teachers. But this is only half of the equation. Less than half, we hazard. Rarely does anyone say, "I will work smarter," although this is the more responsible of the effort twins. Working hard AND working smart are a powerful combination, and it is this "working smart" component that, according to Dweck, is crucial to building a growth mindset.

Cast your mind back to when you were at school. Think of a time when you felt really, really challenged, to the point of feeling overwhelmed with stress. Now, with the insight of age and experience, what are one or two strategies you could have used? Which teachers could you have talked to about helping you find a strategy (they don't necessarily have to be the teacher of that particular class)? How would you know if that strategy you chose to use was working or not? If it was not working, how could you tweak that strategy or what other strategy might you try? Which teacher could you ask for help with this? How we use strategies is important.

These are the steps that Dweck found to be key to building a growth mindset as an iterative process:

1. Use a strategy or strategies.
2. Follow through and actually work hard at this strategy.
3. Evaluate how it works.
4. If it isn't working, if the results are not successful, acknowledge this, do not paper over your (temporary!) failures.
5. But do not stop there. Tweak your strategy or use what you learned to devise a new strategy.
6. Evaluate how this works.
7. Do not do all this in isolation—make sure you are communicating with all of your teachers—as learning happens best in collaboration.
8. Repeat this process. Keep iterating.

Fundamental to this process is that it is actionable and accountable. Building a growth mindset requires both these attributes. Too often, it is the accountability that is missing. Falling short is often met with a platitude along the lines of "at least you tried," which is a world different than "at least you tried that, it seemed like a good idea; now what can we learn from that, what shall we try next?" In the former unsuccessful effort is an endpoint, in the latter it is just a marker in a reiterating process that the student has allies in and control over. This is the difference.

Can one teacher's efforts make a difference? Dweck's research suggests that there is some domain specificity to growth mindset—for instance, a child might have a growth mindset in math but a fixed mindset in art. But the exact boundaries of domain specificity are unclear, and are likely to vary both with the individual and over time.[15] I am reminded of Richard Dawkins' answer to one of the favorite antievolutionary posits: what use is half a wing?

> Half a wing is indeed not as good as a whole wing, but it is certainly better than no wing at all. Half a wing could save your life by easing your fall from a tree of a certain height. And 51 per cent of a wing could save you if you fall from a slightly taller tree. Whatever fraction of a wing you have, there is a fall from which it will save your life where a slightly smaller winglet would not.[16]

What use is a growth mindset in one area? It is certainly better than no growth mindset at all. And who knows what the future will bring, what other teachers and influential adults will intersect with that child's life?

None of us is either 100 percent a fixed mindset person or growth mindset person—our mindsets vary for different demands on our brains and may be different on different days. But one thing we do know is that mindsets can be changed. The cumulative effect of growth mindset appearing in different domains in a child's life might be a true "yet sensibility" that gives a student the confidence to seek, and ultimately tackle, new challenges.

Building a growth mindset has the power to transform lives, and is one of the most precious skills with which we can help equip our students. So the next time a student says, "I will work harder," say yes, that is a good place to start, but I want you to work smarter, too. And begin an enduring discussion with them about what this might mean.

We know that developing a growth mindset or "yet sensibility" in each child is critical as he or she faces those inevitable academic, social, and emotional bumps in the road that all students meet through their academic journey, and that all children face in their lives outside school as they grow up. But some students do not merely face bumps, but rather mountains in their journeys, and this is where research in growth mindset is very promising.

For years, educators and policy makers have been exploring ways for the most disadvantaged groups of students, who often find themselves in some of the most disadvantaged school settings, to experience enhanced academic achievement, to graduate from high school, and to attend college. Research has shown that developing a growth mindset among high-poverty, marginalized groups of students has shown statistically significant improvement in essential skills such as reading.[17]

IT TAKES CHALLENGE

In addition to practice, it also takes *challenge*. Our brains are wired to learn; it is fundamental to being human. But do we make the most of it as teachers? What do you think students will remember from each of the classes they take this year? How would they do if your final exam was actually given at the beginning of the next school year rather than the end of this one; what would they remember? Scary though this prospect is to most teachers, it is arguably a truer test of what has actually been stored in a student's long-term memory. How might we teach with challenge and enduring memory in mind, as goals that are important to us as professional teachers?

If I were to ask any senior at St. Andrew's who he or she interviewed for the nationally recognized Oral History Project they all do as eleventh-graders, there is a strong likelihood that the student would recall not just his or her name, but details of his or her story.[18] But if we asked those same seniors to discuss in detail President Lyndon Johnson's "Great Society," which we know we teach well, it is less likely that they would recall much from this landmark program. But why?

Research shows that when students are challenged, especially by something that interests them, their intrinsic motivation increases and the knowledge and skills are more likely to be imbedded into their long-term memory. However, such projects take time, trial and error, and are often superseded by the need to "cover" material for the short term. Lost along the way is the space for cultivating a "yet sensibility."

One of the pillars of MBE-informed practice is teaching and assessing in multiple modalities, chosen by "what is the best way to teach or assess this?" rather than individual students' perceived "learning strengths." Add to this the idea of using a neurodevelopmental lens to analyze the brain demands germane to the subject you as a teacher know and care about, the brain demands of how you choose to teach it, and the brain demands of how you choose to assess it. How can you bring these more into alignment so that students see an essential authenticity and fairness? Our experience in teaching teachers about MBE science, and one that is backed by research,[19] is that doing so changes how they teach and assess, leads to greater differentiation and greater teacher efficacy, and improves learning for their students. It also creates the moments and a climate for cultivating that "yet sensibility."

We have often dreamed of finding a way to bottle the growth mindset that elementary-school, and to a great extent middle-school, students bring to school each day. If we had such an elixir, we would dispense it to our high-school students and parents whose fixed mindsets take hold as they become blinded by the college process. We have seen this firsthand with our students. In their elementary- and middle-school years, they are free to fail, iterate, and improve, as their grades will never make it onto a college transcript.

But something amazing, or better described as discouraging, happens at the commencement of the ninth-grade year for many students and their parents. They become afraid to make mistakes, or to take intellectual risks, that might have an adverse effect on a grade. Compliance rather than creativity predominates. They become "achievement drones"—*just tell me what I need to do to get an A and I will do it.*

Where the growth mindset of students prevailed, a fixed mindset now takes over. With it, students' self-confidence and sense of well-being tend to become intimately linked with their latest grade, and they begin to label themselves as innately "smart" or "dumb"—and none of these labels of innate ability is good. With terrible irony, Dweck's research suggests that this fixed mindset that is incubated is detrimental to long-term academic and career achievement levels.[20]

But it would be wrong to just blame students and parents—as teachers and school leaders, we are the ones who have a propensity to create achievement-drone assembly lines and factories. In our correct and necessary aim to set a high bar for student learning, we often miss the mark. At this point, let us revisit that earlier notion: what if you moved your final exam to the start of the next school year? What would your students still know? Why is the very notion of doing this preposterous? Given that it sounds preposterous, what does this tell us about the high-school culture we have created? What value does it place on actually learning?

For homework, try this. Find a student, a senior perhaps, and ask him or her about the difference between grades and learning, marvel at his or her eloquence, and cringe at what his or her level of insight means. In a culture of achievement über alles, we appear to have lost sight of the primacy of the goal of learning, and of teaching for learning.

BUT, even if this is not optimally fertile ground for cultivating a "yet sensibility," cultivate it we can and cultivate it we must. We can use the eight-step method outlined earlier in this chapter—the platinum triangle of student–strategies–teacher working in a reflective, iterative manner. We can teach and assess in ways that are more mindful of the balance between long-term reflective learning versus immediate achievement. We might dream of whole schools with a "yet mindset," but any teacher can make it happen in his or her corners of influence, and by doing so, begin to sow the seeds of change.

WHAT DOES A "YET SENSIBILITY" LOOK LIKE?

Providing students models of individuals or teams with a "yet sensibility" often leads us to our passions with sports. As we wrote this book during the summer of 2015, the United States Women's Soccer team was participating and ultimately triumphing in the World Cup of soccer. During the competition, an article appeared titled "These Six US Stars Were Rejected from Youth Teams. It Made Them Great." Such stories abound. Glenn often uses another sporting example to provide models of a growth mindset for my students.

In 1980, the United States and the Soviet Union were pitted against each other in an ideological struggle—the Cold War—that happened to be fought on many fronts, including through sports. Just prior to the Lake Placid Winter Olympic Games, the Soviet professional ice hockey team, without question the best in the world, defeated the United States' team, made up of college players, 10–3 in Madison Square Garden in New York. Thirteen days later, the two teams squared off again in the semifinal game of the Olympics. The US team won 4–3, prompting hall of fame sportscaster Al Michaels to pose the question, "Do you believe in miracles?" While we are not sure of our stance on miracles, we do believe in "yet." Someone was going to beat the Russians, and with a "yet sensibility," those American college students did.

What can a "yet sensibility" look like in the classroom? It begins with earned trust and relationships. Teachers, coaches, and advisors strive to have each of their students meet their peak potential as learners and as individuals. Throughout a student's academic journey, teachers are professionally obligated to help a student not only meet his or her potential but also to see how far they can push that student beyond that initially perceived potential. Part of the earned trust comes from a palpable, unwavering belief in each individual student, to honor the student's *current* strengths and weaknesses, and to work with him or her to demystify the learning experience.

Out of the earned trust comes the platinum triangle of student–strategies–teacher working together, with students trying strategies, evaluating how they are working, tweaking them or choosing new strategies, and doing so in an ongoing discussion with teachers. We feel that a true "yet sensibility" emerges when everyone is doing this—the academic high fliers and the "just fine" students who often get overlooked in grand education debates, as well those students with learning challenges. This works for all students, and having all students do it helps create a "yet sensibility," a growth mindset as part of the school culture.

To help sustain this culture, it is important for teachers and school leaders to model the "yet sensibility" in their own everyday practice—even going out of their way to give public displays of "yet!" at well chosen times. If we preach growth mindset and routinely practice fixed mindset, students will quickly sniff out our hypocrisy.

To help develop effortful, yet-inclined learners, teachers need to define what this is and looks like in a way that is readily digestible by students. At St. Andrew's, we provide students two grades: one for their academic performance in all of their disciplines and one for their effort. Effort grades have been part of our school's mission since its founding, but the process had been a bit muddied. For the most part, students who contributed most during class discussion and did their homework got the highest effort grades. More introverted students were actually being penalized for their effort. Certainly, reading Susan Cain's *Quiet* changed that perception for many of the faculty. So began a journey of looking at the research.

Once again we turned to the work of Dweck as well as the faculty's training in the neurodevelopmental framework of All Kinds of Minds, and identified eight "standards" that we felt are critical to students' academic success and that require deliberate practice: participation, note-taking, materials management, day-to-day learning, self-advocacy, collaborative work, metacognition, and dealing with absences. Within each category, we defined several usually observable criteria, with alternatives for "exceeding expectations," "meeting expectations," "progressing toward expectations," and "not meeting expectations" (table 5.1).

Table 5.1. Excerpt from St. Andrew's Effort Grade Rubric

Standards	Exceeding Expectations	Meeting Expectations	Progressing Toward Expectations	Not Meeting Expectations
Self-Advocacy *Language* *Social Cognition*	☐ Consistently and independently communicates (in-person or electronically) with teacher to schedule extra help, when needed. ☐ Consistently resilient and proactive when faced with a challenge in-class or out.	☐ Usually initiates communication (in-person or electronically) with teacher to request extra help, when needed. ☐ Usually resilient and proactive when faced with a challenge in-class or out.	☐ When extra help is needed, sometimes initiates communication (in-person or electronically) with teacher. ☐ Sometimes resilient and proactive when faced with a challenge in-class or out.	☐ When extra help is needed, rarely initiates communication (in-person or electronically) with teacher. ☐ Rarely resilient and proactive when faced with a challenge in-class or out. Often gives up.
Collaborative Work *Language* *Social Cognition*	☐ Consistently and positively works well with assigned partners/teams. ☐ Consistently and respectfully listens to the thoughts and insights of classmates while sharing their own. ☐ Consistently and independently assists classmates who need support/help.	☐ Usually works positively with most assigned partners/teams. ☐ Usually a respectful listener to the thoughts and insights of classmates. ☐ Usually assists classmates who need support/help.	☐ Sometimes works well with assigned partners/teams. ☐ Sometimes a respectful listener to the thoughts and insights of classmates. ☐ Sometimes assists classmates who need support/help.	☐ Has difficulty working with partners/teams. ☐ Is rarely able or willing to listen to insights and thoughts of classmates. ☐ Rarely (and never independently) assists classmates who need support/help.
Meta-Cognition (Thinking About Learning) *HOC* *Language* *Memory*	☐ Consistently takes advantage of opportunities to reflect (in writing or orally) on academic performance. ☐ Consistently able to articulate current learning strengths and challenges. ☐ Consistently able to explain appropriate personal learning strategies for a given task. ☐ Consistently applies, and is receptive to receiving, teacher's written and oral feedback.	☐ Usually takes advantage of opportunities to reflect (in writing or orally) on academic performance. ☐ Usually able to articulate current learning strengths and challenges. ☐ Usually able to explain appropriate personal learning strategies for a given task. ☐ Usually applies, and is receptive to, teacher's written and oral feedback.	☐ Sometimes takes advantage of opportunities to reflect (in writing or orally) on academic performance. ☐ Sometimes able to articulate current learning strengths and challenges. ☐ Sometimes able to explain appropriate personal learning strategies for a given task. ☐ Sometimes applies teacher's written and oral feedback.	☐ Rarely takes advantage of opportunities to reflect (in writing or orally) on academic performance. ☐ Rarely able to articulate current learning strengths and challenges. ☐ Rarely able to explain appropriate personal learning strategies for a given task. ☐ Rarely applies teacher's written and oral feedback.

But such a rubric is not enough. Reflection is key, and providing students time for "meta-cognitive moments" to reflect on their learning is a critical teaching and learning strategy. Therefore, teachers should intentionally build reflection into their lessons while carefully avoiding "reflection fatigue." There is a fine balance to find here, and finding it is very important.

As an example, one place we have enhanced our use of reflection at St. Andrew's is both before and after assessments of major projects. Prior to a test or project, students are asked what demands this assessment will place on their brains and what skills and strategies they will need to achieve at their highest level. Immediately following an assessment or project, students are asked to reflect briefly on what worked and what did not work, and what, with the benefit of hindsight, they might have done differently.

We have found that there is a brief, fragile, fleeting moment of honesty and vulnerability immediately after an assessment that we can tap into—provided the earned trust of the teacher is there—when many students will reflect on their learning with a remarkable level of insight and honesty. As teachers, we find that taking the time to read these and reply back very promptly (along with the graded assignment) goes a long way to helping forge these important ongoing discussions between the student and teacher about his or her learning, and what roles strategies play in it.

In a meta-analysis of research by the Education Endowment Foundation for their "Teaching and Learning Toolkit,"[21] metacognition, along with peer tutoring and feedback, were deemed to be among the most beneficial and cost-effective learning strategies. We concur. But too much of a thing, even as good as this, can be counterproductive. In order to bring novelty to such reflection, one strategy we have used is to apply many of the excellent "Visible Thinking" routines developed by Ron Ritchhart and his team as part of Project Zero at Harvard University. Ritchhart's work ties in very well with our work on MBE science informed practice.

Creating a "yet" mindset requires teachers to craft assignments and projects that lead students into deep, active cognitive engagement—learning that grips minds. Teachers should strive to create opportunities where students truly care about learning—where the potential joy of learning shouts so loudly it drowns out the background noise of achievement and grades. It is not that these do not matter, it's just that right now they are just not as important as nailing down this really incredible thing my teacher is getting me to do.

We have all created teaching moments like this and can recognize those amazing moments when it happens in a class. Let us make more of them. These are the moments when we can inspire students to work hard and work smart to overcome a monumental "yet" challenge. MBE science research suggests some strategies to help, such as incorporating personal relevancy or choice, or having students work on human-centered problems (empathy is an underutilized tool), as discussed elsewhere in this book. Imagine a school where learning was everywhere, where students encountered and overcame "yet" moments on a regular basis. What would it look like?

Each of us—teachers, students, and school leaders—could as individuals make a difference for the people we love and the world we live in. We just have to give our very best, practice smarter, embrace challenge, trust in our teachers, colleagues, mentors, and parents, and amend each "can't" with a deafening "yet!"

Without looking back from this page, what are the *three* most salient points you take away from this chapter of *Neuroteach*?

What are *two* things you would like to do "tomorrow" with the information you learned from reading this chapter?

What is *one* question you have after reading this chapter?

NOTES

1. See Carol Dweck's online resource at http://mindsetonline.com.
2. Carol Dweck, delivered at *The Sunday Times* Festival of Education, June 19, 2015.

3. Carol Dweck, *Mindsets and Math/Science Achievement*, Carnegie Corporation of New York, Institute for Advanced Study, Commission on Mathematics and Science Education, 2008, http://www.growthmindsetmaths.com/uploads/2/3/7/7/23776169/mind set_and_math_science_achievement_-_nov_2013.pdf, citing research from F. Rheinberg, R. Vollmeyer, and W. Rollett, "Motivation and Action in Self-Regulated Learning," in M. Boekaerts, P. Pintrich, and M. Zeidner, eds., *Handbook of Self-Regulation*, 503–29 (San Diego, CA: Academic Press, 2000).

4. Carl Bialik, "A Lifetime of Career Changes," *Wall Street Journal*, September 3, 2010. See also Carl Bialik, "Seven Careers in a Lifetime? Think Twice, Researchers Say," *Wall Street Journal*, September 4, 2010.

5. Judy A. Willis, "How to Teach Students about the Brain," *Educational Leadership* 67, no. 4 (2009); "Neuroplasticity: Learning Physically Changes the Brain," *Edutopia*, http://www.edutopia.org/neuroscience-brain-based-learning-neuroplasticity (accessed September 19, 2015).

6. See Carol Dweck, "The Power of Believing You Can Improve," TEDTalk, December 17, 2014 and Angela Lee Duckworth, "The Key to Success? Grit," TEDTalk, May 9, 2013.

7. Peter C., Brown, Henry L. Roediger III, and Mark A. McDaniel, *Make It Stick: The Science of Successful Learning* (Cambridge, MA: Belknap Press, 2014).

8. "Brain and Learning," http://www.neafoundation.org/pages/courses.

9. "Neuroscience and the Classroom," http://www.learner.org/resources/series214.html.

10. See chapter 10, "Homework, Sleep, and the Learning Brain."

11. Daniel Coyle, *The Talent Code: Greatness Isn't Born. It's Grown. Here's How* (New York: Random House, 2009).

12. Malcolm Gladwell, *Outliers: The Story of Success* (Boston: Back Bay Books, 2011).

13. Rachel Nuwer, "The 10,000 Hour Rule Is Not Real: The Biggest Meta-Analysis of Research to Date Indicates That Practice Does Not Make Perfect," Smithsonian.com, August 20, 2014.

14. Angela L. Duckworth, Christopher Peterson, Michael D. Matthews, and Dennis R. Kelly, "Grit: Perseverance and Passion for Long-Term Goals," *Journal of Personality and Social Psychology* 92, no. 6 (2007): 1087–101, doi:10.1037/0022-3514.92.6.1087.

15. Carol S. Dweck, *Mindset: The New Psychology of Success* (New York: Random House, 2006).

16. Richard Dawkins, *The God Delusion*, first edition (Boston: Houghton Mifflin Harcourt, 2006).

17. See Dweck, "The Power of Believing You Can Improve" and Eric Jensen, *Teaching with Poverty in Mind: What Being Poor Does to Kids' Brains and What Schools Can Do about It* (Alexandria, VA: ACSD, 2009).

18. See Glenn Whitman, *Dialogue with the Past: Engaging Students and Meeting Standards through Oral History* (Lanham, MD: AltaMira Press, 2004).

19. R. M. JohnBull, M. Hardiman, and L. Rinne, "Professional Development Effects on Teacher Efficacy: Exploring How Knowledge of Neuro- and Cognitive Sciences Changes Beliefs and Practice" (paper presented at the AERA conference, San Francisco, CA, 2013).

20. Dweck, *Mindset: The New Psychology of Success*.

21. "Teaching and Learning Toolkit: The Education Endowment Foundation," https://educationendowmentfoundation.org.uk/toolkit/toolkit-a-z.

6

"MY BEST (RESEARCH-INFORMED) CLASS EVER"

In the following list of words, circle those that you feel you have prior knowledge of:

Missile Gap	Brinksmanship	Berlin Wall
Containment	Peaceful Coexistence	("Iron Curtain")
Domino Theory	Arrogance of Power	Encirclement
Rollback	Third World	Communism
Détente	Satellite Countries	Capitalism
Mutual Assured		Democracy
Destruction (MAD)		

This is how I began my best research-informed class ever. It happened on April 15, 2014. After twenty-one years of teaching, attending professional development conferences, and reading research in the growing field of mind, brain, and education (MBE) science, it all came together. Try this assignment for yourself. Set a timer for two minutes, and off you go.

Like all teachers, I want the content and skills that I teach to stick longer than the next test. I also recognize that teaching is both an art and a science and that each class is like a blank canvas needing to be designed like a Norman Rockwell painting. Research in the field of MBE science informs that design. The goal for *this* class was to introduce key terms of the Cold War and to have students begin applying each term to Cold War events.

Figure 6.1.   Inspired by the painting *Deadline* by Norman Rockwell.

This "best class" lasted eighty minutes and began with the students ar-
riving to class with Billy Joel's "We Didn't Start the Fire" playing. The
students were immediately engaged by this novel moment. While I often
play music in class, it is frequently limited to my New Jersey roots and my
man crush on Bruce Springsteen. But with lyrics such as "Hemingway,
Khrushchev, the Bay of Pigs, Berlin, Begin" playing in the background, the
students began their "Do Now" to circle as many of the Cold War terms
that were familiar to them. This two-minute exercise placed demands on
each student's attention and memory. A "Do Now" is a strategy to immediately engage students with the question, concepts, or theme of the day,
and takes advantage of what some research suggests is one of the two most
opportune times for learning during a class period.

Optimizing the early moments of each class period has been an area of
focus of our school that is informed by research and shaped by the question,
"How effectively do we engage students at the start of each class period?"
In this case, we looked closely at the "Serial Positioning Effect" (more commonly known as the Primacy Recency Effect)[1] in order to maximize each
learning episode. It highlights how knowledge of three aspects of memory

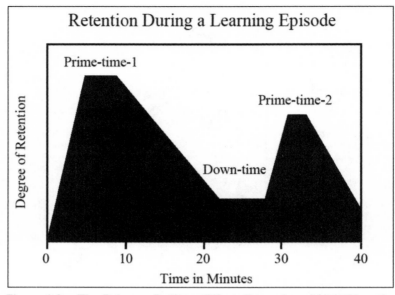

Figure 6.2. The Primacy Recency Effect. Reproduced from *How the Brain Learns* by David A. Sousa, with permission of Corwin Press.

are in demand during every class period: long-term memory, active working memory (sometimes called immediate memory), and short-term memory.

What is important to note is that the first efforts to research this are quite old, dating from the early twentieth century. It is a reminder that research does not have to be brand new to be beneficial to the classroom. However, like most educational research, more needs to be done by teachers to figure out how to use the primacy recency effect to understand how to maximize their time with their students and to get as much of what they teach to stick as long as possible after each class meeting.

Nevertheless, this one "effect" has impacted how many teachers design their classes. They recognize that there is a flow to a student's attention and memory that must be considered in their planning that begins with immediate and relevant engagement, moves to delivery of critical knowledge or skills, then allows time for deliberate practice, and ends with a period of recall and reflection. But exactly what this process looks like for any individual class involves the teacher considering many other factors. Developing the proper flow of a class period takes into account what Vanessa Rodriguez points out in her book, *The Teaching Brain*, that:

> Learning is a dynamic, interactive, context-dependent process. An individual learns distinctly because of her own individual biology (including genetic

inheritances and overall physical health), access to good nutrition, the learning environment, engagement with different kinds of learning tools (books,
websites, diagrams—the list is endless), and teachers. Learners are in constant
interaction with other external influences, such as their friends, family, culture, and society. Together, these internal and external characteristics interact
to form the learning brain.[2]

Back to my class. While a number of the students had some prior knowledge of these Cold War terms, albeit from Hollywood movies, most did not.
In fact, this was the first time most of these students had a chance to dive
deeply into the history of the Cold War.

My students were then introduced to each term through direct instruction, dare we say lecture, an instructional strategy that continues to be critical to student learning. Expert teachers need to know their content really
well and must develop their content knowledge alongside their pedagogical
knowledge, as we discuss further in chapter 12. In order to do higher-order,
critical, and creative thinking, students need knowledge and experiences to
draw upon. And these need to be robust. Therefore, providing a definition
for each of these sixteen Cold War terms was the first step.

The importance of having a foundation of knowledge, and not making too
many assumptions about what students already know, can be seen in research
and has been explored and debated in books such as *Trivium 21c*[3] and *Seven
Myths about Education*.[4] This is why teachers must know really well their
content, and its essential representations and "threshold concepts"[5] critical
for understanding. This targeted use of traditional methods, like lecture,
augmented by MBE informed methods such as formative assessments, can
be effective in building the foundational knowledge students need.

My class then transitioned into small groups that were given twenty
minutes and the use of their laptops to try to connect as many Cold War
terms as possible to Cold War events. Merely putting students into groups
does not ensure learning happens, and laptops have certainly proven to be
effective learning tools and tools of distraction for students (see chapter 11,
"Technology and a Student's Second Brain").

Collaborative learning, or group work, has its place if designed well by
the teacher. It also requires important monitoring by teachers. Too often
in classes we visit, when students are doing group work, teachers are not
engaging with each student enough. They can be seen checking their
email or taking care of other class business rather than moving among the
groups, listening, posing questions, and providing just the right amount
of hints that might aid, in the case of this class, their search for Cold

War events that link to the Cold War terms. It is during these teacher "walkabouts" (as one student has called them) that I often apply some of the Visible Thinking routines presented by Ron Ritchhart and others at Harvard's Project Zero[6] and, for example, ask students "what are they wondering about?" These are great easily implementable routines that can get students thinking in novel, interesting ways.

As the students made connections between the Cold War terms and Cold War events, they were then challenged to find a historical image that correlates to each historical term. While there is some debate over whether it is vision or smell that trumps all other senses,[7] in the learning environment, vision reigns supreme. Therefore, teachers are continually looking for ways to present their material using multiple modalities and engaging multiple senses. This particular multiple-modality activity has a far greater chance to begin the process of embedding these key concepts into each student's long-term memory because students are learning through manipulating information.

Too often, the pressure of covering extensive content means that teachers feel that their responsibility is complete as soon as they tell students what they need to know. However, we know from research that the "empty vessel approach," the "turn on the faucet and fill the heads of students" approach to teaching and learning, shows little evidence of sustained learning. The teacher may feel good that he or she has managed to get through so much "stuff" in so little time, but how much actually stuck? And for how long? Moreover, does the particular teacher actually even care, or do they believe that the moment the content-dump is done, the job of teaching is over, and it is now up to the students to memorize it?

These students may be able to do satisfactorily on the end-of-unit test, but research suggests that storage into the long-term memory—what we might think of as true learning—often does not occur very effectively when this approach is taken. To paraphrase a quotation we will come across later in the book, if teaching is going on but no true enduring learning is taking place, does it actually count as teaching?

One way students can create more enduring learning is by transferring knowledge into new mediums. As a result, one teacher who designs her introduction to Cold War terms similarly to mine, had her students actually draw their own images to support each Cold War term. Dr. Mariale Hardiman's book, *The Brain-Targeted Teaching Model for 21st-Century Schools*,[8] is an important resource for understanding the critical role of knowledge transfer through the arts, across all academic disciplines, as a tool for helping students learn and remember content knowledge.

So how did my class end? Or, rather yet, what is the best way to end a class? Consider this question as you walk down the halls of your school at the end of each class period. Too often, as a teacher is delivering his or her final content or directions to end a class, the students can be seen readying themselves to move on in their day, to their next class or activity. As research suggests, this is a loss of a prime learning moment—one that teachers do not maximize often enough. It is a learning travesty. What should be of more importance to teachers than a solid exit strategy? Without it, much of what students learned during that class period will be lost from their memory.

In the case of my class, students were given an "exit ticket," an opportunity to reflect on and recall on what the instructor wanted them to learn for this class. The concept of an "exit ticket" is not new, and Doug Lemov's book, *Teach Like a Champion 2.0*,[9] affirms its importance as an instructional strategy. We were first introduced to this book through our partnership with the professional development arm of Teach for America in the Washington, DC, region.

At the end of my class, students were handed a note card and asked to list from memory as many of the sixteen Cold War terms as they could recall. The class average was six. This was great feedback for me to help me plan for the next class. The act of recalling helped my students fix in their long-term memory what they at that moment knew, but it also primed their brains to learn more of the sixteen the next time they encountered them. Learning is best when it is iterative.

Recall opportunities (or ROPPs) are important moments that help content stick in students' long-term memory. When we think about how neural connections are made (see chapter 4), "use it or lose it" really is the case. Teaching and learning tend to suffer from a lack of ROPP. Too often, the time-lapse between when teachers introduce a concept and when they assess it is so long that students in fact have to relearn (usually reteaching themselves, often through cram sessions) the material for the short term. Just because many of us studied this way, and maybe we did all right, it does not mean it is the most effective or time-efficient way to learn.

Therefore, when my students returned to class the following day, they were confronted with a "Do Now" in the form of a formative assessment that asked them to once again list from memory as many of the sixteen Cold War terms as they could recall. The new class average was now eight. Periodically throughout the three-week Cold War unit, which on occasion might have been helped by the homework from the night before, this forced recall was required of students.

The rationale behind this was that students needed to master this knowledge content to be able to read primary and secondary sources, to talk at deep levels about the Cold War, and to engage in higher-order thinking tasks that required them to manipulate this knowledge. Even if they didn't get them all correct, it was okay—the act of getting answers wrong is a vital part of the learning process and helps build robust neural pathways. The deconstruction and application of each of these terms thus continued throughout this unit on the Cold War. One research-informed class, even my best class ever, will not ensure long-term memory consolidation. Ultimately it comes down to practice.

As research informed as this class was, the class alone did not ensure that students were able to recall and apply Cold War concepts on the summative assessment at the end of the Cold War unit, or on the final exam at the end of the course. Opportunities to practice, apply, and recall these Cold War terms, spaced out over time, were equally important. Moreover, the need to further practice should not devalue the art of teaching that was integral to the design of this class.

When I first taught the Cold War in 1991, my teaching of the Cold War terms looked a lot different, more "sage on the stage," teacher-directed, "write these down and learn them, class." I continue to believe that having a foundational knowledge of the Cold War terms is necessary—and this idea of deliberately and thoroughly building a solid content knowledge before moving on to discussion and higher-order thinking tasks is relevant to every discipline (see, for example, discussion of the Trivium[10] in chapter 11). But my design of this class has gone through multiple iterations as both my content knowledge and pedagogical knowledge have increased. It is a reminder that being a teacher is a journey, an art, and a science. It is also a reminder of how complex the job of designing your own "best" research-informed class is. Give it a try and give yourself the grand permission of time to make it even better the following year.

Without looking back from this page, what are the *three* most salient points you take away from this chapter of *Neuroteach*?

What are *two* things you would like to do "tomorrow" with the information you learned from reading this chapter?

What is *one* question you have after reading this chapter?

NOTES

1. David A. Sousa, *How the Brain Learns* (Thousand Oaks, CA: Corwin, 2012); Alan D. Castel, "Metacognition and Learning about Primacy and Recency Effects in Free Recall: The Utilization of Intrinsic and Extrinsic Cues When Making Judgments of Learning," *Memory and Cognition* 36, no. 2 (March 2008): 429–37, doi:10.3758/MC.36.2.429.

2. Vanessa Rodriguez, *The Teaching Brain: The Evolutionary Trait at the Heart of Education* (New York: The New Press, 2014), 51.

3. Martin Robinson, *Trivium 21c: Preparing Young People for the Future with Lessons from the Past* (Bancyfelin, UK: Independent Thinking Press, 2013).

4. Daisy Christodoulou, *Seven Myths About Education*, first edition (London and New York: Routledge, 2014).

5. Castel, "Metacognition and Learning about Primacy and Recency Effects in Free Recall."

6. "Visible Thinking," http://www.visiblethinkingpz.org/VisibleThinking_html_files/VisibleThinking1.html (accessed October 6, 2015). See also *Making Thinking Visible: How to Promote Engagement, Understanding, and Independence for All Learners*, first edition (San Francisco: Jossey-Bass, 2011).

7. D. G. McCullough, "The Strange Science behind Our Sense of Smell," *Guardian*, http://www.theguardian.com/sustainable-business/2015/mar/23/smell-fragrance-perfume-nose-aroma-science (accessed October 6, 2015); John Medina, *Brain Rules (Updated and Expanded): 12 Principles for Surviving and Thriving at Work, Home, and School*, second edition (Edmonds, WA: Pear Press, 2014).

8. Mariale M. Hardiman, *The Brain-Targeted Teaching Model for 21st-Century Schools* (Thousand Oaks, CA: Corwin, 2012).

9. Doug Lemov, *Teach Like a Champion 2.0: 62 Techniques That Put Students on the Path to College*, second edition (San Francisco: Jossey-Bass, 2014).

10. Robinson, *Trivium 21c: Preparing Young People for the Future with Lessons from the Past.*

"I LOVE YOUR AMYGDALA!"

Fear prevents the flowering of the mind.

—Juddi Krishnamurti

Teaching is an emotional profession and being a student is an emotional journey. Separating teaching, learning, and emotion from one another is an impossible task and it is something that actually goes against the principles of mind, brain, and education (MBE) science.[1] When teachers and students make an emotional connection to the subject matter, it increases their intrinsic motivation. When students make an emotional connection with the teacher because a teacher believes in them, it increases their self-efficacy. Conversely, when a student is bored, or stressed, or feeling threatened or disconnected, not much learning takes place.[2] So in which direction does a student go on any given day? Surely the teacher must have some role to play?

Emotion is inseparable from learning. The brain is a complex organ, and all cognitive tasks involve multiple parts working together—it is challenging science to master. However, a carefully chosen simplification, which our playful title for this chapter hints at, may aid us in understanding the effects of emotion on the learning brain.

The amygdala, a critical part of the limbic system, may be thought of as the brain's emotional switching station—which means that any teacher who wants his or her students to learn should be deeply invested in "amygdala

Figure 7.1. Fear prevents the flowering of the mind.

management" and mindful of the role of emotion and stress in learning.[3] So, teachers, be bold, one day take a risk and yell out, "I love your amygdala!"[4] and when the puzzled looks rain down on you, you better have an explanation ready. Do not worry, we will help you with this.

Our brains are constantly bombarded by sensory inputs, more than are possible to process, so some filtering is done before incoming information or experiences get to areas like the prefrontal cortex, where executive functioning and higher-order thinking take place, and the hippocampus, where memory tasks begin. In young children, the filtering is not fully developed, hence the great quotation from Alison Gopnik, professor of psychology and philosophy at the University of California–Berkeley, about what it feels like to be inside the brain of a young child, "it's like being in love in Paris for the first time after you've had three double espressos."[5]

The amygdala plays a key role in the brain system that acts as our emotional filter. When we are under stress, it directs sensory intakes to our rear "reactive brain" where our "fight, flight, or freeze" response is embedded, and that automatic response from our evolutionary ancestors kicks in. This is sometimes called "downshifting."[6] Think about a school setting. How much learning takes place at this point? Dr. Mariale

Downshifting

Negative emotions or stress can cause us to lose focus on higher order thinking, and instead process in our brain's emotional center. Think back to a time when you experienced downshifting. Take a few minutes to think about what happened, how you felt, how you reacted. Then write about it below.

Figure 7.2. Downshifting reflection. Image adapted from "Munch—It's a Scream" by Ian Burt, www.flickr.com, February 15, 2006, http://creative commons.org/licenses/by/4.0.

Hardiman writes, "many students in our nation's schools are locked into this downshifted mode of thinking as a result of standard educational practices. Students are thus literally disconnected from their capacity for creativity and learning at high levels. The very institutions charged with developing the creative and higher-level thinking of students are using methods that inhibit this development."[7]

However, when we are under no or low stress, the limbic system that includes the amygdala directs sensory intakes to the prefrontal cortex, home of executive functioning and higher-order thinking. Our thinking, organizing, planning, and problem-solving skills are unleashed on these sensory inputs, so we are primed to learn. This, then, is the first reason why a great teacher must be an amygdala manager. To create conditions where our students' brains are primed to learn, we need to be constantly aware of and carefully balancing stressors and stress levels. Later we will see how a great teacher also helps shape which of these two paths their students' amygdala may go down.

Table 7.1. Factors That Cause and Reduce Stress in Students

Factors that cause stress and lead to the reactive brain fight, flight, freeze response:	Factors that reduce stress and lead to the thinking, reflective brain response:
• Boredom	• Choice
• No personal relevance	• Novelty
• Frustration of previous failures	• Humor, music
• Fear of being wrong if asked to speak in class	• Being told a story or anecdote
• Fear of presenting work orally	• Positive interactions with peers
• Test-taking anxiety	• Acting kindly
• Physical, language, or dress differences	• Movement
• Feeling overwhelmed by workload and unable to organize time to respond to these demands	• Optimism
	• Expressing gratitude
	• Making correct predictions
	• Achieving challenges

While lots of teachers probably want to guess what many of the factors that lead to high stress and low stress are, MBE science provides us with research to form our list. Table 7.1 shows the results of fMRI (functional magnetic resonance imaging) studies.[8]

Picture a student sitting in class or at the dining room table staring at his or her homework, his or her amygdala being bombarded by all these sensory intakes that, just possibly, could lead to learning. What could the teacher do to limit the chances of all this good stuff just heading to the brain's fight, flight, freeze zone of no learning? Some of this comes from assignments that are set.

Actively work to limit the deadening, unnecessary, and very real stress of boredom because it is a terrible, unconscionable waste to shut down minds that would otherwise be engaged and motivated. Find ways to add relevance to students' lives, which increases engagement and motivation. Use low-stakes formative assessments to help students see what they know and what they do not know, and to give the teacher feedback to guide their future teaching, both of which can help reduce the frustration of previous failures and test-taking anxiety.

Some of the stress reduction comes from classroom structures and scaffolds put into place that aim to reduce some of these fears and frustrations. This important work requires the continuing reflective effort of teachers and schools. It is part of the classroom community-building power that approaches such as Reggio Emilia[9] and Responsive Classroom[10] instill.

Part of it comes from recognizing that time and materials management are such crucial factors to student success that we should not just assume they will grow optimally on their own.

Elementary-school teachers know that it is important to explicitly nurture these skills, but as students get older we tend to leave them to fend for themselves. But these executive functioning skills stem from the prefrontal cortex, an area of the brain that retains neuroplasticity all the way through college. So we have the potential to help students rewire their brains to be better at these amazingly important skills that will help them throughout school, college, and beyond. But it gets better. Research also shows that developing executive functioning skills may also help our children's developing brains to manage cortisol levels, making them better able to cope with stress, even toxic stress,[11] when it does occur.[12]

But a major part of reducing unnecessary stressors is balancing strategies that might be implemented in attempts to reduce the stress caused by boredom, with factors that may actually increase stress—and recognizing that this balance point is different for every student with whom the teacher interacts and is shifting continually. Given how busy, complex, demanding, or stressful students' lives outside school tend to be, the teacher is trying to support each student without a full set of data. As we will learn later, there is even more to balancing the stress see-saw than this, but doing so is crucially important if we want learning to take place.

To help with this balancing act, and to help create classes where learning takes place, the second list is full of what Dr. Judy Willis calls "dopamine boosters," factors that increase dopamine levels and help send sensory inputs toward a student's thinking, reflective brain.[13] Maybe the teacher gives students a choice (or perhaps the carefully crafted illusion of choice). Maybe class starts with a short, entertaining, maybe seemingly unconnected video clip that you masterfully weave into your lesson at some point. Maybe students walk into a room filled with music. Maybe students see visual evidence of their own recent thinking captured on the walls of the classroom. Maybe the teacher helps engineer genuine moments of laughter.

Perhaps we use stories. We know that storytelling is powerful, one of the pieces in the core of what it means to be human, but as students move to higher and higher grades we tend to put storytelling aside. Don't! Storytelling, and its equally powerful sibling, storyfinding, which taps into a student's all too rarely used powers of empathy, where he or she is now suddenly working for a person who has cares, thoughts, feelings, and needs, can be transformative.

Again, some well-respected teaching approaches such as Reggio Emilia and Responsive Classroom have many of these factors at their core, which is part of why they work. In addition, the "Visible Thinking" work of Ron Ritchhart at Harvard's Project Zero[14] provides good tools to engage stu-

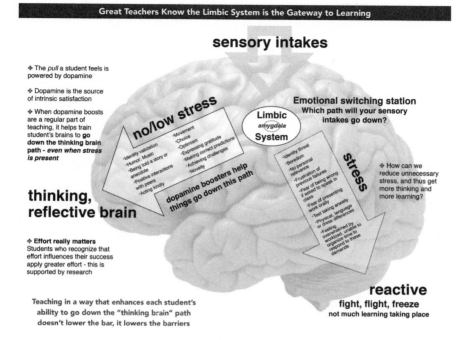

Great Teachers Know the Limbic System is the Gateway to Learning

sensory intakes

✦ The *pull* a student feels is powered by dopamine

✦ Dopamine is the source of intrinsic satisfaction

✦ When dopamine boosts are a regular part of teaching, it helps train student's brains to **go down the thinking brain path** - *even when stress is present*

thinking, reflective brain

✦ **Effort really matters** Students who recognize that effort influences their success apply greater effort - this is supported by research

no/low stress
•Identity validation
•Humor; Music
•Being told a story or anecdote
•Positive interactions with peers
•Acting kindly
•Movement
•Choice
•Optimism
•Expressing gratitude
•Making correct predictions
•Achieving challenges
•Novelty

dopamine boosters help things go down this path

Limbic *amygdala* System

Emotional switching station
Which path will your sensory intakes go down?

stress
•Identify threat
•Boredom
•No personal relevance
•Frustration of previous failures
•Fear of being wrong if asked to speak in class
•Fear of presenting work orally
•Test taking anxiety
•Physical, language or dress differences
•Feeling overwhelmed by workload, unable to organize time to respond to these demands

✦ How can we reduce unnecessary stress, and thus get more thinking and more learning?

reactive
fight, flight, freeze
not much learning taking place

Teaching in a way that enhances each student's ability to go down the "thinking brain" path doesn't lower the bar, it lowers the barriers

Figure 7.3.   A simplified diagram showing the brain's response to high- and low-stress situations in the classroom.

dents of all ages in deep thinking in ways which are more likely to provoke a "reflective brain" response.

Stress affects memory, too. The amygdala's reaction to stress appears to play a pivotal role in helping us store emotionally significant memories.[15] However, stress inhibits the everyday long-term memory formation that is crucial to learning. Cortisol produced in the brain by stress causes the hippocampus not to function as well—and it is the hippocampus that converts the contents of our short-term memory into long-term memory storage. This is worsened by the fact that most people overestimate, or rather misidentify, what short-term memory is.

Research suggests typical limits of short-term memory in adults of seven items and thirty seconds[16] (but not both! the more items you need to hold in short-term memory, the less time you can do it for), with children able to store fewer items for less time. Thinking about a typical class, this means that long-term memory storage is involved much more frequently than you might have thought. Stress impedes memory storage, which impedes learning.

The final impact of stress may seem disjointed in our discussion, as it does not fit the strategies we will discuss next, but we include it because of its enormity. Research suggests that long-term exposure to high levels of stress, particularly in early childhood where much of the brain's stress-response system is developing, can lead to a decrease in cognitive ability and memory function, and can lead to a range of mental and physical health problems, including suppression of the immune system, cardiovascular disease, diabetes, stroke, depression, anxiety, and alcohol and drug problems—all of which can last way beyond the duration of the stress.[17]

While much of the toxic stress to which children are exposed comes from factors outside the school, schools cannot abdicate responsibility. The stakes in terms of lifelong health and learning, and thus the future, are too high. Teachers and schools have a duty not to add unnecessary stress, but recognizing the plasticity of the brain during school years, particularly the stress-response system in early childhood, schools are uniquely positioned to be agents of positive difference. As Jack Shonkoff says, "better outcomes are likely when neural circuitry is 'wired properly' from the beginning."

MBE science tells us that a typical, unstressed, doing-the-work student might manage to store just 50 percent of what he or she is told by teachers in long-term memory. So what about a student who chronically suffers from high levels of stress, if life at home is not so good, or he or she is being bullied at school? What percentage is he or she learning? This is a physiological response, not something the child chooses, so until we find structures and supports to help the stress situation, other interventions we might do to aid learning are like building a house on weak, crumbly foundations. Now think of that child's lifelong health, and that child's lifelong learning ability too. Addressing early childhood chronic stress is a hard mission for a nation's schools, but it is a three-for-one win.

So what should the "I love your amygdala teacher!" do? First, find ways to reduce or eliminate unnecessary stressors that, at their core, are barriers to learning. But be aware, we often fall into the trap of confusing *creating barriers to learning* with *maintaining rigor*, and we urge the reflective teacher to always hold this in their mind. The power of equipping teachers with a neurodevelopmental lens, such as that provided by All Kinds of Mind, which defines a hierarchy of mind and brain skills, helps here, too. It allows teachers to analyze the neurodevelopmental demands germane to their discipline, the actual thinking skills inherent in the subject the teacher is passionate about, and come to some alignment among these skills, the skills demanded by how you teach, and the skills demanded by how you assess.

Out of this three-pronged alignment comes a kind of academic fairness that students see and appreciate. It helps teachers to place a greater variety of thinking demands on a student, all of which ring true to the discipline or topic, and helps them add deep challenge in authentic ways. So we do now hereby give permission to the passionate, reflective teacher: look deeply into the heart of the subject you care about . . . what does rigor really look like here? What will you do to align the thinking skills entwined in your picture of rigor with those you demand of your students when you teach and those you demand when you assess? The goal of the "I love your amygdala!" teacher then, as Dr. Willis herself says, is to "lower the barriers, not the bar."

Second, the "I love your amygdala teacher!" should use strategies, such as those listed in table 7.1, that help the amygdala choose the prefrontal cortex reflective brain path. The added benefit to this is that the more the young developing brain does this, the more myelinated this pathway gets—the more likely the reflective brain will be utilized in stressful situations. The concept of neuroplasticity extends to the school's ability to help students develop more resilient minds—a powerful prospect, but this takes mindful teaching. Research from MBE science gives us some additional strategies to develop resilience, as we explain next.

Third, research suggests that well-developed self-organization, executive-functioning abilities can help children deal more effectively with episodes of stress, so developing modes of teaching and assessment to deliberately foster these skills is beneficial.[18] The plasticity of the prefrontal cortex, the key area of the brain for this, into early adult years means that time spent helping students develop ever more robust organizing and planning skills is good at all levels of schooling rather than just adopting a "they should be good at this by now" attitude.

At this point, hold a picture of a student in your mind—now imagine him or her working through medical school or law school, picture him or her working in a demanding, fulfilling career of your choice. How does having a robust set of organizing and planning skills, and confidence in this, help him or her here? This is far, far from wasted work as a teacher.

Fourth, research from Jack Shonkoff suggests that a certain degree of "tolerable" stress may actually aid children's ability to deal with future stress—as long as the stress occurs in a supportive environment where children knows they have safe, dependable relationships to which they can turn, and the stress events are not too long in duration, too intense, or too frequent. This then gives great teachers three additional tasks:

1. Identify the kinds of teaching, projects, and assignments that create what Shonkoff calls, "low levels of manageable adversity [which] have been shown to serve as a form of 'stress inoculation' that can enhance later resilience." Surely including factors such as relevance, choice, purposefully helping the growth of executive functioning skills, and being mindful of the variety and authenticity of neurodevelopmental skills demanded, is important here, too?

2. Encourage teachers and school leaders to create environments where every child feels safe, heard, and connected, and where all students know they have safe, dependable relationships to which they can turn. Ron Ritchhart uses the word "enculturation" to encompass all the factors in a school that add up to create a feeling that students get that the school is a place where thinking takes place, and where they can be a thinker as part of this community.

Our wonderful colleague Rodney Glasgow, founder of the National Diversity Directors Institute, added the following two crucial factors to our stress lists in table 7.1: the role of identity threat in causing stress, and the role of identity validation in reducing stress. This inexorably links all the work that the spheres of education are currently doing around ideas of diversity, equity and inclusion, and community with the work of MBE science; this underexplored link must and will grow. The impact of violence and bullying in all its forms on stress and learning must be included as well, since emotion and learning are inseparable (and the effects of chronic stress, as noted above, are so significant).

The 2014 study conducted by St. Andrew's Episcopal School and researchers from the Harvard Graduate School of Education that lead Research Schools International[19] found that happiness at St. Andrew's correlates with intrinsic motivation, but not extrinsic motivation—that the joy came from the worthiness of the task itself rather than from external rewards. It also found that happiness correlates with academic achievement, with students' responses suggesting that factors such as relationships with teachers and administrators were almost as important to them as relationships with peers, and that both were major factors contributing to happiness and thus intrinsic motivation and academic performance.

Of course, this did not happen by accident, but we highlight it as an example of what is possible when attention is paid to the environment that is created, the types of teaching and learning that take place, and the types of professional development that happen.[20]

3. Heighten the importance of the "I love your amygdala!" teacher's ability to balance the stress see-saw: not too much, but not too little. The stress seesaw, of course, varies from student to student and from moment to moment. So the teacher must ensure that each individual student is in what we might call his or her zone of proximal discomfort—for *this* student at *this* time, what is a good level of stress and what can I do to help balance his or her stressors?

This concept of the zone of proximal discomfort can be shown overlaid on David Diamond's reworking of the curve created by psychologists Robert Yerkes and John Dodson to describe the observed relationship between arousal (which may include tension, anxiety, stress) and performance, as shown in the diagram below, which we include primarily to help teachers visualize the task of balancing the stress see-saw.[21] Research tells us that nothing is more important to enhancing student learning than the quality of the teacher.[22]

And this should be no surprise even if we looked at the task of balancing the stress see-saw alone, keeping every student in the zone of proximal discomfort. Think of a modern fighter jet that is inherently unstable and wants to fall out of the air. But its electronic brain reads so many millions of data a second from the environment around it and makes so many millions of calculations and decisions a second based on what it learns that the whole thing functions and stays in the air. This is the "I love your amygdala!" teacher.

Last, perhaps the most important task and the inspiration for the title of this chapter, is the simplest to do. Talk to students about this, show them the diagram, explain to them the role of their limbic system and that because of neuroplasticity, they have the ability to rewire their brain to make themselves more resilient and better learners. Such a dialogue will further demystify learning and all of its complexity. Then, help them on this journey.

Most schools have philosophies that are remarkably similar, both in content, with words like "community," "challenge," "responsibility," and "lifelong love of learning," and in the fact that few if any employees actually know what they are. What if your school's philosophy was "We make learning fun." That, we know, sounds jarringly fluffy. Maybe you picture classrooms with a soft-serve ice cream machine in the corner. But let's unpack it.

For learning to be fun, it needs to be challenging. Easy quickly becomes boring and is most definitely not fun. When students trust the adults around

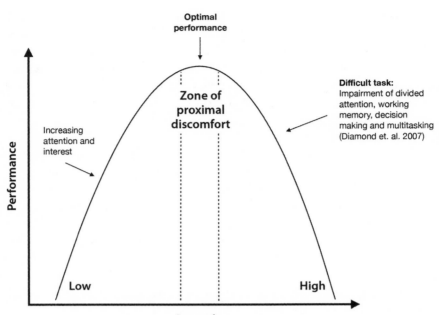

Figure 7.4. The zone of proximal discomfort and the Yerkes-Dodson curve. The Yerkes-Dodson curve has been the subject of significant debate since its creation in 1908. Whilst it does not appear to hold true for all cognitive tasks, Diamond et al. (2007) suggest that it holds true for "complex" tasks, ones that involve the prefrontal cortex. Since this is the home of executive functioning and higher order thinking, this would include the majority of tasks that teachers set students—and particularly the ones that cause them stress. Adapted from David M. Diamond et al., "The Temporal Dynamics Model of Emotional Memory Processing: A Synthesis on the Neurobiological Basis of Stress-Induced Amnesia, Flashbulb and Traumatic Memories, and the Yerkes-Dodson Law." *Neural Plasticity* (2007). Based on the original work of R. Yerkes and J. Dodson, "The Relation of Strength of Stimulus to Rapidity of Habit Formation." *J. Comp. Neurol. & Psy.* 18 (1908): 459–82.

them, and trust their peers because school leaders and teachers have worked to create a culture in which all students feel known and respected for who they are, learning can reach that crucial level of being more challenging because students are more likely to take a risk: a risk with a thought, an action, or just the risk of being invested.

It is fun for students when they themselves know they are doing something significantly challenging—but it is transformative because the task is well crafted to be intrinsically motivating and they trust the community that surrounds them. Fun speaks to challenge and community, fun speaks to excellent teachers, fun speaks to the kind of schools where significant learning takes place; the kind of schools where stress occurs, but where the stress see-saw is carefully managed by teachers who love their students' amygdalae.

Without looking back from this page, what are the *three* most salient points you take away from this chapter of *Neuroteach*?

What are *two* things you would like to do "tomorrow" with the information you learned from reading this chapter?

What is *one* question you have after reading this chapter?

NOTES

1. Christina Hinton, Kurt Fischer, and Catherine Glennon, "Mind, Brain, and Education: The Student at the Center Series." *Mind, Brain, and Education*, March 2012, http://www.jff.org/sites/default/files/publications/materials/Mind%20Brain%20 EducationPDF.pdf.

2. David A. Sousa, *How the Brain Learns* (Thousand Oaks, CA: Corwin, 2012); "Neuroscience and the Classroom: Making Connections," funded by Annenberg Learner, produced by Harvard-Smithsonian Center for Astrophysics Science Media Group, 2011, retrieved from www.learner.org/courses/neuroscience; Hinton, Fischer, and Glennon, "Mind, Brain, and Education: The Student at the Center Series."

3. Judy A. Willis, "What You Should Know about Your Brain," *Educational Leadership* 67, no. 4 (2009), retrieved from http://www.ascd.org/ASCD/pdf/journals/ed_lead/ el200912_willis.pdf.

4. Although the brain is a complex organ and the amygdala is just part of the stress-response system, this is a useful phrase because it is catchy enough to linger in people's minds, reminding them to be mindful of the role of emotion and stress in learning.

5. Alison Gopnik, *The Philosophical Baby: What Children's Minds Tell Us about Truth, Love, and the Meaning of Life*, reprint edition (New York: Picador, 2010); "Alison Gopnik: What Do Babies Think?" TED.com, October 3, 2014, https://www.ted.com/ talks/alison_gopnik_what_do_babies_think?language=en.

6. Geoffrey Caine and Renate Nummela Caine, *The Brain, Education, and the Competitive Edge* (Lanham, MD: Rowman and Littlefield Education, 2001).

7. Mariale M. Hardiman, *The Brain-Targeted Teaching Model for 21st-Century Schools* (Thousand Oaks, CA: Corwin, 2012).

8. Judy Willis, presentation to St. Andrew's Episcopal School (Maryland) faculty, 2013.

9. Reggio Emilia is an approach to primarily preschool and elementary education, named after the town in Italy where it was developed by Loris Malaguzzi. "A child has a hundred languages," able to express their thoughts and show their understanding in many different ways, is a guiding phrase. Children investigate, explore, and construct their own learning, with teachers as mentors and guides, learning with the children.

10. Responsive Classroom is a research-based approach to education that focuses on four areas: (1) teachers create a positive community where all students feel comfortable; (2) teachers create a calm, orderly environment that promotes autonomy; (3) teachers continually use observations to make the class developmentally appropriate; (4) teachers create engaging activities by making them appropriately challenging, interactive, and related to children's interests.

11. Jack Shonkoff defines toxic stress as follows: "Toxic stress refers to strong, frequent, and/or prolonged activation of the body's stress-response systems in the absence of the buffering protection of adult support. Major risk factors include extreme poverty, recurrent physical and/or emotional abuse, chronic neglect, severe maternal depression, parental substance abuse, and family violence. The defining characteristic of toxic stress is that it disrupts brain architecture, affects other organ systems, and leads to stress-management systems that establish relatively lower thresholds for responsiveness that persist throughout life, thereby increasing the risk of stress-related disease and cognitive impairment well into the adult years."

12. J. P. Shonkoff, "Leveraging the Biology of Adversity to Address the Roots of Disparities in Health and Development," *Proceedings of the National Academy of Sciences* 109, Supplement 2 (2012): 17302–307; J. P. Shonkoff, W. T. Boyce, and B. S. McEwen, "Neuroscience, Molecular Biology, and the Childhood Roots of Health Disparities: Building a New Framework for Health Promotion and Disease Prevention," *JAMA* 301 (2009): 2252–59.

13. Judy Willis, *Research-Based Strategies to Ignite Student Learning: Insights from a Neurologist and Classroom Teacher*, first edition (Alexandria, VA: Association for Supervision and Curriculum Development, 2007).

14. Ron Ritchhart, Mark Church, and Karin Morrison, *Making Thinking Visible: How to Promote Engagement, Understanding, and Independence for All Learners*, first edition (San Francisco: Jossey-Bass, 2011).

15. Willis, "What You Should Know about Your Brain."

16. Nelson Cowan, "What Are the Differences between Long-Term, Short-Term, and Working Memory?" *Progress in Brain Research* 169 (2008): 323–38. *PubMed Central* (accessed October 3, 2014).

17. Shonkoff, Boyce, and McEwen "Neuroscience, Molecular Biology, and the Childhood Roots of Health Disparities"; Shonkoff, "Leveraging the Biology of Adversity to Address the Roots of Disparities in Health and Development."

18. Shonkoff, "Leveraging the Biology of Adversity to Address the Roots of Disparities in Health and Development."

19. In the spring of 2013, The Center for Transformative Teaching and Learning at St. Andrew's Episcopal School was the eighth school in the world to be invited to become part of Research Schools International that is led by researchers at Harvard University's Graduate School of Education (HGSE). Research schools are thought-leaders in education that bridge the gap between research and practice. This is a global network of schools in the United States, Europe, Australia, and South America that conduct cutting-edge research, lead professional development, and disseminate research findings to the broader educational community in partnership with faculty from HGSE.

20. Lauren Schiller and Christina Hinton, "Happier Students Get Higher Grades in School: Research Study Says," *TIME*, August 7, 2015, http://time.com/3984782/happy-students-high-grades/.

21. David M. Diamond, Adam M. Campbell, Collin R. Park, Joshua Halonen, and Phillip R. Zoladz, "The Temporal Dynamics Model of Emotional Memory Processing: A Synthesis on the Neurobiological Basis of Stress-Induced Amnesia, Flashbulb and Traumatic Memories, and the Yerkes-Dodson Law," *Neural Plasticity* 2007, doi: 10.1155/2007/60803, based on the original work of R. Yerkes and J. Dodson, "The Relation of Strength of Stimulus to Rapidity of Habit Formation," *Journal of Comparative Neurology and Psychology* 18 (1908): 459–82.

22. John Hattie, "Teachers Make a Difference: What Is the Research Evidence?" (paper for Australian Council for Educational Research Annual Conference, 2003); John Hattie, *Visible Learning: A Synthesis of Over 800 Meta-Analyses Relating to Achievement*, first edition (London and New York: Routledge, 2008).

8

MEMORY + ATTENTION + ENGAGEMENT = LEARNING

Human beings, who are almost unique in having the ability to learn from the experience of others, are also remarkable for their apparent disinclination to do so.

—Douglas Adams

Prior knowledge. The ability to recall, apply, and connect it is essential for deep and enduring learning. Tapping it is an essential part of solving meaningful problems. But how do we effectively and efficiently help students build their own personal body of prior knowledge?

At a workshop we gave at a prestigious independent school in the United States, we learned that the previous year the science department had the fabulous idea to replace their traditional end-of-year exams with an exam in September at the start of the following school year. The students knew this was coming, knew the exam counted as a significant grade, but the results were, as the Science Department had guessed, spectacularly bad. We suspect that the only "one-off" part of this story is the courage of the aforementioned teachers to do it, and that the same result would hold true in most classes in most schools. How many teachers have heard some version of the phrase, "I learn it for the test and then immediately forget it." There is a not-too-subtle clue here that something is wrong with "learning" as we tend to witness and perpetrate it.

What is learning? Courtney Clark and Robert Bjork, cognitive psychologists at the University of California–Los Angeles, give us an important distinction that helps us reframe the nature of teaching: the difference between "learning" and "accessibility."[1] A major problem in education is that we like to think we are promoting learning when actually a lot of schooling is centered on promoting and assessing accessibility.

Accessibility is reflected in a student's current performance, but does not necessarily indicate that underlying long-term learning has or has not taken place. Clark and Bjork argue that high current accessibility—being able to recall recently acquired facts—does not necessarily mean that "learning" has taken place. Accessibility is fragile, so the student may not be able to recall this information later: enduring long-term learning has not necessarily taken place. Clark and Bjork argue that the opposite is also true—just because a student's current accessibility is low, it does not mean that learning has not occurred: "students can be learning even when a current test shows little or no progress . . . In sum, when we seek long-term learning, current accessibility is not a reliable index of how effective a lesson has been."

This idea that current accessibility is not a reliable index of learning has profound implications in how we craft the flow of a school year, how we assess learning (by which we mean the enduring long-term learning of Clark and Bjork), and how we coach students to study.

The role of teachers is to build to assess learning, but the current trend of assigning a grade to everything that moves suggests that a lot of what they are doing is assessing accessibility—the current level of retrieval. Once we understand that this is different than long-term learning, we should shift our practice, both in promoting methods that increase enduring long-term learning and in creating assessments designed to measure enduring long-term learning rather than accessibility. The latter, we believe, includes freeing students from the endless litany of high stakes for-a-grade quizzing, a topic we have already discussed, arguing that this practice should be replaced by low-stakes formative quizzes, the main job of which is to give both teacher and student feedback about what steps to take next.

This recognizes that learning, true learning, may take a while, and we need to give students space in order to do it. This is particularly true because research suggests that getting things wrong is a key part of memory formation, and the introduction of what Clark and Bjork call "desirable difficulties" that lead to students making errors is a more effective way of creating enduring learning than errorless learning.

But since our new goal as teachers is to create "errorful" learning, we need to rethink how, or if, we grade these necessary errorful steps. Maybe we give a low value grade based on effort, maybe we give no grade at all. Whatever we do, we must find ways to keep students in the game while they experience being wrong on the journey to enduring learning. Given that some students may be more often wrong than others along the journey to the same ultimate level of learning, it seems unfair to punitively punish the journey.

The good news is that research from mind, brain, and education (MBE) science has suggestions for errorful strategies to developing enduring learning, such as spacing, interleaving, and testing. Layered onto these are other strategies for creating engagement, which also help develop learning. In chapter 9 we also discuss assessments that measure enduring learning. But first, let's think of how a typical student might study.

Many students study for a test by rereading and rereading their textbook or class notes, often with the application of rainbow highlighters. But this just creates an illusion of learning. It keeps current accessibility high, but does not increase enduring learning. The familiarity this method breeds gives the students a false sense of security. This rereading often occurs in a massed frenzy of studying a day or two before the test. Again, this aids accessibility rather than learning, but that is okay because, both in the students' minds and often in reality, they only have to know the information for twenty-four hours, then they can happily let it slip from their minds.

Robert Bjork and Elizabeth Bjork[2] argue that it is hard for a student to increase underlying learning when his or her level of accessibility is high. The key to increasing learning is "desirable difficulty"[3]:

> If accessibility can be reduced, however, learners can make bigger gains in underlying learning. Introducing difficulties into learning—thereby reducing accessibility—removes the learner from the experience of having the to-be-learned material at hand. Overcoming the difficulty through subsequent practice is then more effective and can create a new, higher level of learning.

This is what spacing, interleaving, and testing do. Let's discuss these in more depth.

SPACING

Students need to be directed to use spaced studying rather than massed studying. Doing so aids long-term memory consolidation. Eventually students may do this independently, having discovered that this strategy really does work, but until they reach this point, the teacher needs to help them. This means equipping students with scaffolded study guides, review opportunities, and opportunities to get feedback on their reviewing, all in plenty of time so that they can space their review.

They also need to be given time to space their review—if students are given nightly homework assignments of reading, writing, problem sets, and worksheets, when are they going to do the spaced review they need? Then it is no wonder that reviewing is crammed in with a few days to go. Teachers need to schedule what they value.

The other point to mention again here is that the prefrontal cortex— the part of the brain responsible for executive functioning tasks such as planning ahead and executing that plan—is not "fully" developed until a student's mid-twenties. Until this time, significant neuroplasticity exists, and some degree of neuroplasticity continues for the rest of the student's life. This means that (1) our students are currently not as good at organizing as they will ultimately be; (2) this is not really their fault; (3) we have the opportunity and responsibility of helping them get to this point; (4) we can both have high standards and help students scaffold these essential learning tasks. Instead of being at odds, these two positions are actually utterly reconcilable.

INTERLEAVING

This is the practice of mixing in questions and topics from earlier in the year throughout the rest of the year. Or, to put it another way, do a bit of this, then a bit of that, then a bit of this again. This aids long-term memory consolidation, and leads to enduring learning. For all that teachers complain that their students have a "one and done" approach to learning material for tests, we tend to promote it by having a "one and done" attitude to tests.

Teachers, therefore, need to mentally readjust the flow of the school year to keep circling around to revisit old material, or to structure the year to do a "bit of this and a bit of that," moving between topics that are constantly building on each other. Material from earlier in the year is to be expected on assessments, with the teacher offering scaffolded review, removing the scaffolding by layers and with judgment as the year goes on. If this makes assessments more challenging, so be it. It is the right thing to do to create enduring learning, and we have to trust the professional skills of teachers still to grade this new paradigm fairly.

TESTING (INCLUDING ACTIVE RETRIEVAL AND PRETESTING AND FORMULATIVE QUIZZES)

Research suggests that the act of trying to recall information aids memory formation. Testing, then, is a very important tool to create enduring learning. Teachers typically see it the wrong way round, that testing is solely the means to assess learning. It is that, too, but its most powerful effect is to create enduring learning. In this regard, it can be used in many different ways.

Self-Testing

In the parlance of the classic flash card, the task of staring at the blank side trying hard to recall is the most crucial step, even if you are unable to remember the information. Then checking to see if you are right and adjusting your thinking are important, but successful learning all stems from that act of trying to recall. We believe that many students get this the wrong way round, and turn the card over too quickly to read when they initially draw a mental blank—and that this is indicative of how students see self-testing in general.

Active study methods that rely on self-testing, such as taking a blank sheet and trying to write down everything you know on a subject, making flash cards (and using them correctly, with the mental emphasis on recall rather than reading!), attempting review questions, or trying to tell someone all you know about a subject, work better than passive methods, such as rereading notes or the textbook. Rereading creates the illusion of mastery—words and phrases become familiar which leads students to believe they know the material better than they actually do. Compared to rereading, self-testing will probably feel more difficult and more mentally draining for students, but research and our experience suggest it will create significantly more enduring learning. It is one of the most powerful strategies in this chapter.

Pretesting

Research suggests that starting a unit of study with a pretest helps create more enduring learning. It appears to give students something on which to hang subsequent information. This test should, of course, not be graded, or if it is, it should be graded for effort rather than correctness. The other point of this pretest is to give the teacher an idea of where the level of the class generally is, and what knowledge each individual student has already, so that the teacher can tailor subsequent classes to best match the needs of the class.

It is important to avoid seeding boredom, and to avoid the potential skipping of foundational knowledge that could prevent future learning. These are two common toxic effects on learning. In addition, students notice and appreciate this attempt at responsiveness, which aids engagement and motivation.

Formative Assessments

This is the contemporary reincarnation of the pop quiz. Research suggests that frequent testing aids memory consolidation and recall, but these should be formative assessments, either for no grade, a low grade, or graded on effort. "Formative" means they help form memory, rather than judge it. Regular use of formative assessments and pretesting creates an important shift in students' minds about the purpose of such assessments—they realize that their purpose really is to help them learn.

Students get honest and frequent feedback about what they know and what they do not know—regular feedback as to where their level of expertise *currently* is. They can use this to guide their future studying: What topics need more studying? What do they need to see a teacher about because they really don't know it? What should they not "waste" much more precious time on because they know it really well already?

There is also a metacognitive aspect that students could be prompted to include: What study strategies appear to be helping? What strategies do not seem to be working? Furthermore, what can be done about this, what other strategies can be tried, and who can help?

Students learn and appreciate that the goal of formative assessments is to prepare them for a later summative assessment—there is a cyclical nature to learning of trial, feedback, retrial, feedback, and so on, with the goal that learning will indeed happen through the patient application of effort. This is different from the standard for-a-grade quiz model, so typical in schools, that incentivizes a mindset that learning has to be immediate; and if it is not, you are, in varying degrees, a failure.

The other benefit of formative assessment is the feedback it gives the teacher about how both the class and individual students are doing so that they can alter their teaching or pacing accordingly. Again, like pretesting, this shift to a more visibly responsive teaching approach helps increase student engagement and motivation. Using formative assessments instead of pop quizzes or high-stakes quizzes is another one of the most powerful strategies in this chapter.

Research from MBE science suggests even more strategies for increasing how much students will remember, such as the arts integration and the primacy recency effect.

THE PRIMACY RECENCY EFFECT

Students will remember best what occurs at the beginning of the class. They will remember second best what happened at the end, as discussed in chapter 6 and shown in figure 6.2. Learning can happen in the middle, and this is far from wasted time, but it takes more effort. So do not waste the precious opening part of the class doing routine activities, like checking homework.

The middle time might be good for having students use newly acquired knowledge in some way. The end time is precious and often undervalued—so don't rush, rush, rush until the bell, physical or metaphorical, sounds. Plan it knowing that this is one of the key moments where more students have a better chance of remembering what you say. In the words of Stephen Colbert, "Exit strategy: have one."

ARTS INTEGRATION

Research suggests that arts integration, using the arts as an everyday teaching methodology in other non-arts areas of the curriculum, helps create enduring learning, and also increases student motivation.[4] This is also a way to craft assessments that incentivize enduring learning rather than accessibility. It is an idea that intersects with another key MBE research-informed strategy, teaching and assessing in multiple modalities.

Learning in this fashion may feel more difficult for the student than aiming for the low bar of temporary accessibility. But research suggests that the payoff for this effort is that the resulting learning will be more enduring. The teacher has an important role to play in coaching students in these methods, with the goal of having these be the students' go-to strategies when they study independently. It is the teacher's responsibility to get students to deeply believe that these are the methods they should use. Therefore:

- The teacher needs to provide assessments that incentivize enduring learning and disincentivize accessibility. This is discussed more in chapter 9, "Assessment 360°."
- Teachers have a role in educating students about why they are teaching and assessing in this way—arming students with small, well-chosen nuggets of MBE science can go a long way.
- The teacher needs to actively coach strategies for enduring learning, and do so all the way through high school (remember, the prefrontal cortex, which plays an important role here, remains significantly plastic all the way through college, so we have a responsibility to create an environment and experiences to help students maximize their ability).

The teacher must do this for the high-flying students, the "just fine" students, and those with learning challenges—these strategies will work for all students. For example, if we can help students reclaim just one hour of sleep each night by creating memory storage more efficiently, that will be a major victory. Remember, don't confuse "more difficult" with "takes longer"—these learning strategies may feel more mentally demanding to do, but may actually take less overall time to get the same degree of learning.

In addition, self-testing should help students realize when further intense studying is either not necessary (because they have formed enduring memory) or futile (they need to talk to the teacher to gain more understanding)—both of which should save time.

- The teacher also has a role in providing students with opportunities for reflection on how different strategies they are using are working for them. A few questions at the end of a test is one effective method of doing this, capturing a moment of honesty that often comes from the moment of vulnerability that a test creates.
- Teachers and schools have a role in educating parents in how best to support students at home.

If learning this way feels more demanding for students, it highlights the massive need for teachers to purposefully generate engagement. Too often we teachers seem to believe that the subjects we teach are just so awesome that anyone in their right mind should just be able to see that and launch with gusto into the work we set.

Some students will, and those who do not, well, our subjects are just so awesome that they should. In fact, the intrinsic awesomeness of our subjects means that it does not particularly matter how we deliver them, the awesomeness will shine through and drive the "good" students who "get it" to do well, which brings us to perhaps the most memorable pastiche of teaching in a movie: *Ferris Bueller's Day Off*. The scenario we have just painted may seem comic, but we argue that too much teaching in nonelementary grades is currently too much like this and Ben Stein's "Bueller? Bueller? Bueller?" teaching, as shown in the easy-to-find clip on YouTube.

Research from MBE science tells us that without active cognitive engagement, learning does not happen.[5] (We believe there is a transatlantic language barrier here that has caused some prior confusion: the word "engagement" alone in the United States tends to be used to mean what in the United Kingdom might be referred to as deep, active cognitive engagement. In the United Kingdom, "being engaged" at school tends to mean just giving the appearance of being busy with something.)

The real power of purposefully creating deep, active cognitive engagement becomes apparent when we examine the concept of neuroplasticity. Research tells us that there is a significant genetic component to ability, but layered onto this is the fact that our brains continuously adapt based on the environments in and with which they operate and interact. Our experiences in all the places where we spend time, and all the activities we do, act continuously to sculpt our brain architecture. This concept is called neuroplasticity. There is enough research evidence to put this on the list of "science facts." And the prime age range when this happens coincides with our school years—all of them.

Learning reorganizes the connections between neurons in certain areas of the brain. Each neuron is highly connected to others, but when we learn, only some of these connections will be activated, others will not. Over time the connections that are activated more often will be strengthened while those that are not will be weakened or pared away—"use it or lose it." Thus the learning experiences we have shape the organization of our neurons—this is believed to be the basis of memory.

The cumulative effect of experiences over time on the organization and connectivity of neurons is what we might call learning. Whether they acknowledge it or not, teachers and schools play a significant role in this neural reorganizing, because they are in charge of a significant fraction of experiences that a child has—and we do not just mean in formal learning settings. Therefore, students' abilities are not fixed, despite the genetic component of ability, but rather are continually developing, shaped by the learning experiences they have. This development helps shape their academic achievement. The goal, therefore, is for teachers, school leaders, and parents to have a positive influence on how this happens.

Fundamental to this is active engagement. Research suggests that without active engagement, this process of neural reorganizing and strengthening, this critical neuroplasticity, does not take place.[6] As Christina Hinton, Kurt Fischer, and Catherine Glennon put it, *"In educational terms, this suggests that passively sitting in a classroom hearing a teacher lecture will not necessarily lead to learning."* Or, in the words of Ben Stein, *"Anyone? Anyone?"*[7]

Engagement is fundamental to learning. Therefore, teachers are necessarily in the business of creating engagement. Research from MBE science suggests ways to do this.

- Work to ensure that your students feel heard, listened to, and known.

    Research suggests that teachers being cognizant of the emotional needs of each individual student is important for developing motivation and is a critical component of learning.[8] Furthermore, research from The National Longitudinal Study of Adolescent Health has shown a strong association between school connectedness and every risk behavior studied.

- Design your curriculum so that students will see meaning, relevancy, or emotional connections to their own lives. Actively work at helping students make these connections.

    Research suggests that doing so increases attention, engagement, and learning outcomes.[9]

- Actively work both to provide students with experiences and environments that aid identity validation, and to eliminate factors that may cause or enhance identity threat.

    Research suggests that identity threat creates toxic stress that inhibits learning. Conversely, identity validation helps increase student engagement and motivation, leading to better learning outcomes.[10]

- Students need to know that "effort matters most" and the brain has the ability to rewire itself (neuroplasticity). Coach your students that deliberate effort can rewire the brain and lead to enhanced academic achievement.

    Research suggests that students with a growth mindset, "who believe they can become more intelligent by working hard to learn,"[11] do just that.[12] "If a student believes that intelligence is mostly a matter of effort, that student is more likely to be motivated to exert effort, attempt difficult academic tasks, and persist despite setbacks, confusion, and failure."[13] Associating success with their effort, not smartness, is key to helping students learn to persist.[14]

- Students need to know the anatomy of their brains, especially the roles of the prefrontal cortex, amygdala, and hippocampus in learning. Teach your students about this.

   Teaching students the mechanism behind how the brain operates, and teaching them approaches they can use to work that mechanism more effectively helps students believe they can create a more intelligent, creative, and powerful brain.[15]

- Provide students with specific learning objectives.

   Doing so puts students in charge of their own learning.[16]

- What does your classroom look like? Is it stimulating but not over-stimulating? Does it change? Do you show recent student work that changes to keep up to date? Do you create novelty? Are the key objectives of your class prominently displayed at the front of your class?

   Research shows that a stimulating and changing classroom appearance correlates to attention and engagement.[17]

- Teach and assess in multiple modalities. Vary the modality to better suit the content, not to match individual students' learning styles.

   Doing so leads to increased engagement and deeper understanding, and consolidates learning in more powerful ways than traditional testing.[18]

- When you return work, do so quickly, and provide scaffold feedback, allowing students time to struggle to correct their errors, rather than simply marking it right or wrong or giving them the answer.

   Research shows that timely feedback is shown to increase engagement, deepen students' memory of the material, and improve achievement.[19] Scaffolding feedback leads to better retention than does simply providing answers or allowing students to generate answers until they reach the correct one.[20] Allowing students to redo work increases engagement, motivation, and achievement.[21]

- Students need more opportunities to reflect and think metacognitively on their learning strategies and performance. Build in moments for reflection. For example, before each of your assessments, have students reflect on how they will study and what demands a particular assessment will have on their brain. At the end of each of your assessments, require students to answer a few reflection questions concerning how they studied, how effective their methods were, or what they think their grade might be.

> Research suggests that reflection aids metacognition, and is linked to increased motivation, engagement, long-term memory storage, and performance. Structured reflection activities help students grow their own personal metacognition toolkit, which helps empower them to be more independent learners.[22] Teachers can help by providing explicit instruction on metacognition strategies and by modeling metacognition themselves out loud in their own classes. Reflection and metacognition can be used effectively even in early grade levels. One additional reflection method is to use exit tickets in the last few minutes of class, requiring students to recall information from the class and to pose a question such as, "What question do you still have after class today?" Journaling or portfolio assessment are also possible ways to incorporate reflection.

- Offer students choice in subject matter and/or assessment.

  > Offering students choice correlates well to engagement and achievement.[23] Furthermore, "students with a mastery mindset . . . are more likely to choose more difficult but rewarding ways to demonstrate learning."[24]

- Students need opportunities to transfer their knowledge through the visual and performing arts, so integrate art into the core content of non-arts subjects to enhance learning.

  > Arts integration creates knowledge transfer by forcing students to synthesize and rearrange ideas, and is linked to long-term memory consolidation and engagement.[25]

- Use play constructively, in age-appropriate ways, to help build student engagement.

  > Play is critical to the social, emotional, and intellectual development of every child.[26]

We think that the most striking thing about this list of strategies is that they are essentially simple. Research from MBE science will continue to inform more, but even if these are the only strategies teachers work at implementing, learning will be enhanced because engagement—deep, active cognitive engagement—will be enhanced.

Remember, neuroplasticity means that the nature of the learning environment affects how students' brains are rewired. This, then, is a simple list that can make a powerful difference in student achievement—and will do for all students: the high fliers and the "just fine" students, as well as those with learning challenges. But it also goes beyond achievement, as these are strategies to create more efficient, confident, and independent learners.

What about the list of memory strategies to increase enduring learning? Those are simple, too, and will also help create higher achieving, more efficient, confident, and independent learners. Memory + attention + engagement = learning. This, then, is our entranceway to a new definition of teaching excellence informed by research from the field of MBE science. We hope you see that this entranceway, full of simple strategies, some of which are already held as important by teachers, is eminently approachable. Perhaps, we would argue, to the point where avoiding them is unconscionable.

There is one final point to reiterate, and that is the importance of emotional connectedness for creating engagement and intrinsic motivation, a topic explored more in chapter 7, "I Love Your Amygdala!" Research from MBE science once and for all settles any debate—emotion is intimately linked with learning. We know this from our emerging understanding of how the brain works.[27] There is an emotional gateway to enduring learning, and teachers, whether they acknowledge it or not, play a role making sure that gateway is open.

However much we think the subjects we teach matter, however awesome the lesson we have planned is, it is all for nothing—enduring learning will not happen—if the emotional connectedness of the students is not there. There is a truly significant difference between "teaching our subjects to our students" and "teaching our subjects," and it means teaching in the moment, every moment, for each individual in the class.

A NOTE ABOUT ATTENTION

The title of this chapter is a riff on a common phrase, "attention + memory = learning," which we have avoided, in part, because of misconceptions about attention, and in part because creating deep, active cognitive engagement is the important thing for the teacher to do.

Who owns the responsibility for a student being attentive? The teacher. Someone who was at one of our workshops said that his mother, who was also a teacher, had told him that discipline was 90 percent curriculum. The brain is always attending to something, but it just might not be what the teacher wants—we argue that good teaching is about switching the probability in your favor.

In popular usage we think of attention as a matter of respect and self-control, but in neuroscience attention refers to a semiautomatic brain system that helps people manage the flow of information entering their brains, and is not necessarily something they can control. Our brains are constantly being bombarded by far more information from all our senses than they can possibly process, and our attention system helps control this flow so that there isn't chaos. It helps allocate the brain's limited processing resources. The attention system reduces our sensitivity to distracting information and enhances our sensitivity to important information.

People, of course, vary widely in their abilities in each of these aspects (and from sense to sense). For example, some people might be good at filtering out distracting information but not good at enhancing their sensitivity to important information. Such a student might sit respectfully still in your class, following your every movement, but not necessarily taking in much of what you say.

Conversely, consider a student who is not good at filtering out distracting information, so he or she might be bouncing around a bit in class, but might still be good at enhancing his or her sensitivity to important information, so even though this student appears distracted, he or she may actually be taking in what you are saying. The upshot is, just because the common usage of the word "attention" is happening in your class, with lots of respectful and polite behavior, it doesn't automatically follow that the neuroscience usage of the word "attention," which is crucial for enduring learning, is happening throughout the room.

The way to get that to happen is to use teaching and assessment methodologies that focus on creating engagement. The classic pattern of lecture, lecture, quiz, lecture, lecture, test to a room of upturned polite faces may work for some students, but not all, and it is largely due to differences in their brains' attention systems that, while it is subject to some neuroplasticity, is beyond their control in that there isn't a mental switch they can flick to make them learn. This classic pattern, repeated throughout the year, may give the illusion of good teaching, but we argue that it is not good teaching. There is a place for it, but as a part of a much richer palette of everyday methodologies. We argue that engagement—deep, active cognitive engagement—not attention, is the key.

Without looking back from this page, what are the *three* most salient points you take away from this chapter of *Neuroteach*?

What are *two* things you would like to do "tomorrow" with the information you learned from reading this chapter?

What is *one* question you have after reading this chapter?

NOTES

1. C. M. Clark and R. A. Bjork, "When and Why Introducing Difficulties and Errors Can Enhance Instruction," in V. A. Benassi, C. E. Overson, and C. M. Hakala, eds., *Applying the Science of Learning in Education: Infusing Psychological Science into the*

Curriculum (Washington, DC: Society for the Teaching of Psychology, 2014), retrieved from http://teachpsych.org/ebooks/asle2014/index.php.

2. R. A. Bjork and E. L. Bjork, "A New Theory of Disuse and an Old Theory of Stimulus Fluctuation," in A. Healy, S. Kosslyn, and R. Shiffrin, eds., *From Learning Processes to Cognitive Processes: Essays in Honor of William K. Estes*, vol. 2, 35–67 (Hillsdale, NJ: Erlbaum, 1992).

3. Clark and Bjork, "When and Why Introducing Difficulties and Errors Can Enhance Instruction."

4. Luke Rinne, Emma Gregory, Julia Yarmolinskaya, and Mariale Hardiman, "Why Arts Integration Improves Long-Term Retention of Content," *Mind, Brain, and Education* 5, no. 2 (2011): 89–96.

5. Christina Hinton, Kurt Fischer, and Catherine Glennon, "Mind, Brain, and Education: The Student at the Center Series," *Mind, Brain, and Education*, March 2012, http://www.studentsatthecenter.org/sites/scl.dl-dev.com/files/Mind%20Brain%20 Education.pdf (accessed March 2014).

6. Hinton, Fischer, and Glennon, "Mind, Brain, and Education: The Student at the Center Series."

7. Hinton, Fischer, and Glennon, "Mind, Brain, and Education: The Student at the Center Series."

8. "Neuroscience and the Classroom: Making Connections," funded by Annenberg Learner, produced by Harvard-Smithsonian Center for Astrophysics Science Media Group, 2011, retrieved from www.learner.org/courses/neuroscience; P. Graviano, Rachael D. Reavis, Susan P. Keane, and Susan D. Calkins, "The Role of Emotion Regulation and Children's Early Academic Success," *Journal of School Psychology* 45, no. 1 (2007): 3–19; Daniel Willingham, *Why Don't Students Like School? A Cognitive Scientist Answers Questions About How the Mind Works and What It Means for the Classroom* (San Francisco: Jossey-Bass, 2009); Mariale Hardiman, *The Brain-Targeted Teaching Model for 21st-Century Schools* (Thousand Oaks, CA: Corwin, 2012).

9. David A. Sousa, *How the Brain Learns* (Thousand Oaks, CA: Corwin, 2012).

10. Geoffrey L. Cohen and Julio Garcia, "Identity, Belonging, and Achievement: A Model," *Interventions, Implications, Current Directions in Psychological Science* 17 (2008): 365; T. Schmader, M. Johns, and C. Forbes, "An Integrated Process Model of Stereotype Threat Effects on Performance," *Psychological Review* 115 (2008): 336–56; J. Aronson, ed., *Improving Academic Achievement: Impact of Psychological Factors on Education* (San Diego, CA: Academic Press, 2002).

11. Hinton, Fischer, and Glennon, "Mind, Brain, and Education: The Student at the Center Series."

12. Carol S. Dweck, *Mindset: The New Psychology of Success* (New York: Random House, 2006).

13. Carol S. Dweck, *Self-Theories: Their Role in Motivation, Personality, and Development (Essays in Social Psychology)* (Philadelphia: Psychology Press, 1999).

14. Jennifer A. Mangels, Brady Butterfield, Justin Lamb, Catherine Good, and Carol S. Dweck, "Why Do Beliefs about Intelligence Influence Learning Success? A Social Cognitive Neuroscience Model," *Social Cognitive and Affective Neuroscience* 1, no. 2 (2006): 75–86.

15. Judy A. Willis, "How to Teach Students about the Brain," *Educational Leadership* 67, no. 4 (2009); Judy A. Willis, "What You Should Know about Your Brain," *Edu-*

cational Leadership 67, no. 4 (2009), retrieved from http://www.ascd.org/ASCD/pdf/journals/ed_lead/el200912_willis.pdf.

16. Donna Wilson and Marcus Conyers, *Five Big Ideas for Effective Teaching: Connecting Mind, Brain, and Education Research to Classroom Practice* (New York: Teachers College Press, 2013).

17. Mariale Hardiman, *The Brain-Targeted Teaching Model for 21st-Century Schools* (Thousand Oaks, CA: Corwin, 2012).

18. Jeffrey D. Karpicke and Janell R. Blunt, "Retrieval Practice Produces More Learning Than Elaborative Studying with Concept Mapping," *Science* 331, no. 6018 (2011): 772–75, doi:10.1126/science.1199327; Mariale Hardiman and Glenn Whitman, "Assessment and the Learning Brain: What the Research Tells Us," *Independent School Magazine* (Winter 2014).

19. Harold Pashler, Nicholas J. Cepeda, John T. Wixted, and Doug Rohrer "When Does Feedback Facilitate Learning of Words?" *Journal of Experimental Psychology, Learning, Memory and Cognition* 31, no. 1 (2005): 3–8; Keri L. Kettle and Gerald Häubl, "Motivation by Anticipation: Expecting Rapid Feedback Enhances Performance," *Psychological Science* 21, no. 4 (2010): 545–47.

20. B. Finn and J. Metcalfe, "Scaffolding Feedback to Maximize Long-Term Error Correction," *Memory and Cognition* 38 (2010): 951–61; Lisa K. Fazio, Barbie J. Huelser, Aaron Johnson, and Elizabeth J. Marsh, "Receiving Right/Wrong Feedback: Consequences for Learning," *Memory* 18, no. 3 (2010): 335–50.

21. Hardiman and Whitman, "Assessment and the Learning Brain: What the Research Tells Us."

22. Donna Wilson and Marcus Conyers, *Five Big Ideas for Effective Teaching: Connecting Mind, Brain, and Education Research to Classroom Practice* (New York: Teachers College Press, 2013).

23. Hardiman, *The Brain-Targeted Teaching Model for 21st-Century Schools*; Willingham, *Why Don't Students Like School?*; P. M. Miller, "Theories of Adolescent Development," in J. Worell and F. Danner, eds., *The Adolescent as Decision-Maker* (San Diego, CA: Academic Press, 1989).

24. Hardiman and Whitman, "Assessment and the Learning Brain: What the Research Tells Us."

25. Luke Rinne, Emma Gregory, Julia Yarmolinskaya, and Mariale Hardiman, "Why Arts Integration Improves Long-Term Retention of Content," *Mind, Brain, and Education* 5, no. 2 (2011): 89–96; Hardiman, *The Brain-Targeted Teaching Model for 21st-Century Schools*.

26. Stuart Brown, *Play: How It Shapes the Brain, Opens the Imagination, and Invigorates the Soul* (New York: Avery Trade, 2010); Willingham, *Why Don't Students Like School?*; L. S. Vygotsky, *Mind in Society: The Development of Higher Mental Processes* (Cambridge, MA: Harvard University Press, 1978).

27. Hinton, Fischer, and Glennon, "Mind, Brain, and Education: The Student at the Center Series."

9

ASSESSMENT 360°

A simple theory of learning: Learning happens when people have to think hard.

—Professor Rob Coe

Do your assessments, and all the preparation leading up to them, and anything you might do after them, cause your students to think hard? Are they challenging because they force, or maybe even cajole, your students to think hard, or because they consume lots of their time? If it is somewhere in between, where on the continuum between these two points would you place it?

A former history teacher at Blair Academy (New Jersey), where Glenn once worked, prohibited his students from referring to a summative assessment as a "test" or "quiz." Instead, he required his students to describe such assessments of their historical understanding as "learning opportunities" (LOPPs). This deliberate word choice was a fun way to reduce the inevitable stress that students develop when faced with tests and quizzes that would impact their grade and, at least in their minds, their college admission.

Glenn adopted a similar vocabulary for his students but for different reasons. If designed correctly, and with research in mind, assessments are, in fact, prime learning and long-term memory-consolidation opportunities. But this requires educators to think more holistically about assessments and

Figure 9.1. **"Fair" assessment. A famous education cartoon shows a dog, a seal, a fish, an elephant, a penguin, a monkey, and a bird standing in front of a teacher's desk, with the teacher saying, "For a fair selection everybody has to take the same exam: please climb that tree." This is our take on it.**

to look at ways in which research should inform how students prepare for a LOPP, take a LOPP, and reflect on their LOPP performance.

Students often continue to choose "cramming" as their preferred study strategy, then immediately put that assessment out of their minds as they move on to the next one. We now know from research that there is a better way. It also requires educators to think differently and deeply about what a LOPP can look like. We now know from research that there is great room for thoughtful exploration.

Like most teachers, we love our disciplines and want what we teach to stick beyond the summative assessment that ends a unit or is part of a state-mandated test. Teachers tend to pour great passion into the preparation of their lessons because of this. But when we began our teaching careers, we likely fell back on ways in which we had been taught, perhaps assuming our own teaching heroes were infallible. For example, we may have found ourselves giving out review sheets with fifty unit terms two days before a test and leaving it up to the students to figure out how to study them, realizing that only a handful would be on the test.

After the test was finished, we graded them, correcting those wrong answers on the test for the students, and giving them back to the students before moving on to the next unit. Or, we might have given students a surprise quiz that can paralyze some students' ability to share what they know and understand regardless if they were prepared. As current research now tells us, these strategies were not a recipe for long-term memory consolidation. Yes, it may have worked some of the time for some students—including us, perhaps—but this is not the high-percentage game.

There is no question that teachers need to know what students know and don't know about the factual and procedural knowledge of each class they take.[1] We strongly believe that it is with such foundational knowledge and skills, acquired by students in each stage of their learning, that they are able to do more creative and higher-order thinking tasks in future contexts. Assessing what students know and can do is important for gauging where the class should go next, as well as valuable in and of itself. The content and skills students are expected to know can be mandated by national and state standards or individual teachers and school leaders who have more autonomy over their curriculum.

But if you really want to see how innovative a school is, inquire about its thinking and practices regarding assessment. For the students, does the mere thought of assessment trigger stress? Do the teachers rely heavily on high-stakes, multiple-choice, bell-curve-generating tests? Or do the students seem relaxed and engaged as teachers experiment with new forms of assessment designed to support deep and lasting learning? The former may remind us of the overbeatified assessments of our own schooling in the days before the great body of mind, brain, and education (MBE) research was available to us. The latter hints that there might be an alternative, more effective and efficient, way forward.

Too often, teachers think too narrowly when it comes to assessments and leave students to figure out how best to learn what they need to know. We argue that how students prepare for an assessment and reflect on their performance are prime, but often overlooked, learning opportunities. At times, the coaching students are given is limited to merely how to take the test. Moreover, there is often too much distance between when an idea and skill are introduced and when students are forced to retrieve and use that information.

Without building self-reflection into their progress, teachers and students miss a prime learning opportunity to learn from mistakes and successes. So this chapter offers a new rhythm to how teachers should think about assessments that includes six critical elements: (a) more frequent formative (low- or no-stakes) assessments; (b) deliberate practice by each student using active recall techniques; (c) pre- and post-assessment reflec-

Assessment 360°

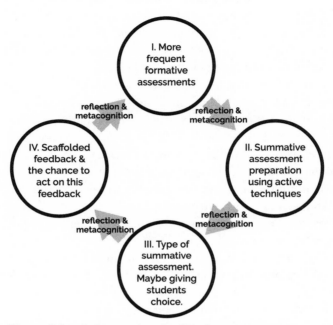

Figure 9.2.   A four-point rhythm for assessment.

tion and metacognition; (d) summative assessment design that includes
a variety of brain demands germane to the discipline, and which might,
at appropriate times, include student choice; (e) scaffolded feedback and
the chance to act on this feedback; (f) a circular or spiraling structure for
content that links old information to new. You will notice that for some the
teacher is the direct agent of action, and for others their role is to influence
the actions of their students.

Clearly, there continues to be a place for traditional assessments of learning that prepare students for standardized tests such as ERBs, AP exams,
SATs, college midterm exams, or PISA. However, we believe that teachers
and school leaders, informed by research, can take on an important leadership role in transforming K–12 assessments from a mere measure of short-
term learning to a crucial component of the teaching and learning process.
In this system of assessment, learning becomes personalized and adaptive,
encouraging students to adopt a mindset focused on discovery and engagement rather than grades and test scores. Most important, in such a system,
the sharp lines between teaching, learning, and testing diminish.

This shift in thinking about assessment is a central issue for all schools. As the research strongly suggests, when students focus on mastery of learning rather than on their performance on tests, they significantly increase their intrinsic motivation for learning.

New forms of assessment can be found in an increasing number of schools today. One example is in a tenth-grade United States history class where we teach. Instead of offering students a typical final exam, they are offered a choice via the question: "What type of final exam would help you best to demonstrate what you have learned in this class?" Based on their reflection, students select the format of their final exam. The more traditional and familiar option allows students to sit for two hours and complete an array of multiple-choice, document-analysis, historical-geography, and free-response questions. The alternative option is to complete the "Historian's Head"[2]—a collage of images and scholarly narratives that respond to the essential questions of the course.

In any given year, 60 to 80 percent of the students choose the assessment that asks them to demonstrate their understanding through images and scholarly narratives—even though this alternative is more demanding when it comes to higher-order thinking and executive functioning skills. It is also more time intensive. So why do so many students choose this option? Because it is novel, visual, and self-directed.

Here are what a few representative students have said:

I function better on projects than on tests because I do not work well under time constraints. In a two-hour exam, I get too stressed and I take too long on one answer. With the project, I am not timed and therefore I will not be stressed.

My primary reason for creating a "head" over a traditional exam is because it will allow me to be creative and use my knowledge in a different way besides memorizing what I have to know. Also I will be able to dedicate more time to it than I would to studying for an exam.

I chose this option over the regular exam because it has an artistic element to it and we are able to spend more time on it that enables a process of deeper thinking in contrast to the quick memorization that would be necessary for the traditional final.

The journey to designing this assessment has been long. Not surprising, when the teacher who designed this assessment began his teaching career in 1991, his assessments looked a lot like those he experienced as a student. While instinct and intuition, and student feedback, suggested that there must be a better way to measure student understanding, ultimately the growing body of research in educational neuroscience became the game-changer for this tenth-grade history final exam.

What we do know is that a growing volume of assessment-related research has shed light not just on the importance of students' mindsets, but also on the importance of continual feedback and how active retrieval of information (memorization recall), in carefully spaced intervals, can produce long-lasting learning. Research also shows that providing students with assessment choice enhances attention and engagement—and confirms that the arts can help deepen long-term memory consolidation.[3] These latter points were instrumental in driving the changes to this tenth-grade history final exam. Specifically, the following research is helping to change our understanding of the correlation between teaching and learning—and altering our approaches to student assessment.

MINDSETS: PERFORMANCE VERSUS MASTERY GOALS

Research on the connection between motivation and learning has focused on two types of mindsets that students develop, based on the kind of experiences (including assessments) that we present them with in school. Students tend to develop either *performance-related goals* or *mastery goals*.[4]

Performance-related goals are those that are linked to more traditional types of assessments. Students become motivated by the grades they achieve, their rankings compared with other students, and extrinsic rewards such as honor rolls or school awards. In contrast, students who develop *mastery goals* are motivated by the actual learning experiences. Their rewards arise from the challenges of acquiring and applying new knowledge and skills.

While students may possess a combination of both types of goals, those motivated primarily by performance goals tend to lose motivation and confidence when faced with difficult academic challenges or when set back by failures. In contrast, students who are motivated by mastery goals are more likely to persevere in the face of such challenges. Difficult tasks or setbacks do not diminish their motivation or self-esteem.[5] Students with a mastery mindset, similar to those students who described their experiences in the history class at St. Andrew's, are more likely to choose more difficult but rewarding ways to demonstrate learning.

FEEDBACK ON PERFORMANCE

Another focal point of research is the timeliness and method of providing feedback to students. Timely feedback has been shown to deepen one's memory for the material assessed.[6] Moreover, in a study comparing immediate versus delayed feedback, the mere anticipation of timely feedback produced better performance. Students who knew that they would get immediate feedback performed better on a task than those who were told that feedback would be delayed.[7]

Just pause to think that through again—the students didn't actually receive any feedback, or additional teaching, they just *thought* they would get rapid feedback, and that made a difference. Finally, studies suggest that marking answers right or wrong (as in multiple-choice tests) has little effect on learning. However, providing the correct response only after a student has spent time struggling to find the correct answer significantly increases retention of the material.[8]

Researchers have also discovered that the spacing of assessments and feedback produces more effective learning. If students revisit content over carefully spaced intervals, they retain information longer than if presented with information once and then only assessed immediately after initial (short-term) mastery.[9]

These findings, among others, show us the reciprocal relationship between assessment and learning. Providing timely and effective feedback can improve student mastery of the content and produce more efficient—and satisfying—learning experiences.

ACTIVE RETRIEVAL OF INFORMATION

A growing body of research suggests that actively retrieving information by self-testing produces significant long-term benefits for learning compared with passive studying, such as merely reading one's class notes.[10] While any assessment requires some type of active retrieval, having students reconstruct what they know through alternative assessments leads to deeper understanding and consolidates learning in more powerful ways than traditional testing.[11]

Active retrieval is also crucial as students prepare for assessments. Too often, when we ask students to reflect on their study strategies, they say they simply "reread class notes." Research tells us that this is not an effective or time-efficient strategy. Training students to build regular self-testing into their study strategies will help them embed material into their long-term memory.[12]

The act of trying to recall something aids long-term memory storage and recovery. By long-term memory, neuroscientists mean, and teachers should be aiming for, not just the end-of-unit or even end-of-semester test, or even the end of year final exam. When we talk about long-term memory, what we mean is, what endures beyond this? Getting to this point takes, as we discuss in chapter 5, both hard work and smart work.

As we like to say to our students, they must develop deliberate study habits and we recognize that "one needs motivation in order to enter into and sustain the hard work of deliberate practice. But the learning happens not simply through putting in the hours, but through doing so intelligently."[13] We feel that it is the role of the teacher to help students with this, embedding research-informed strategies at well-timed moments alongside content, and not just leaving learning to chance. One example is to provide students with recall opportunities while coaching them on how to be good at this strategy. Since studying this way may *feel harder* at times for students, there is an element of emotional support here, too.

FROM THEORY TO PRACTICE

Moving neuroscientific research on assessment into classroom practice faces two barriers. First, despite the fact that the work of educators targets the organ of learning, the brain, most teachers and school leaders have little understanding of the architecture of the brain and how it receives, filters, and applies information. Second, work that falls under the various versions of "brain-based" is too often believed to be of greatest benefit to struggling learners, those who historically might be called learning-challenged. Such thinking is shortsighted and fails to recognize the importance of research in educational neuroscience for *all* students—including the academic high fliers and the "just fine" students who often get ignored in discussions on improving learning. One of our main aims for this book is to begin to address these two issues.

Moreover, it seems that too many parents do not want to believe what K–12 teachers intuitively know: that *every* student needs help and guidance to meet his or her peak potential and that there should be multiple pathways to learning what teachers, schools, and state or national associations have deemed what students should know at various developmental levels.

A parent of an Advanced Placement History student posed the question, "What can educational neuroscience do for my son? He is already getting all A's." This question came at a time in the school year when hundreds of thousands of students were gearing up for AP exams and/or final course exams. These cumulative assessments of learning place intense demands and stress on each student—stress that, research tells us, inhibits optimal learning and demonstration of understanding.

Our response to this AP student's parent came in the form of a question: "What if I told you that the lessons from educational neuroscience research could potentially reduce the number of hours your son spends studying?" This question should pique any parent's interest. Research highlights how crucial sleep is for both learning and health—could five hours studying be reduced to three hours, for example? Could we create room for students to have more time to pursue nonclassroom activities that bring them balance and joy? Could we create more time for them to work at maintaining the positive relationships that we know help underpin learning? Could we help them reduce the depth or duration of their school-induced stress?

It would take some combination of studying efficiently and effectively using strategies informed by MBE research, and the teacher using MBE research to set better homework, as we discuss in chapter 10. One interesting student strategy that we suspect would make a difference, with which we have had a hard time making progress, is multitasking, research on which challenges the way most students study today—especially the transaction cost of switching between social media and academic responsibilities.

What teachers assess should be what they want students to know, not just for an upcoming test, but also for the long term. This places added importance on the relationship between the teacher and their content, as discussed in the "instructional core" section of chapter 12. How we assess students has important implications for whether students will actually retain that content or procedural knowledge for the short or long term.

Far too often, as neuropsychology professor Tracey Tokuhama-Espinosa points out, "While students manage to keep enough dates, facts, and formulas in their head to pass the test, this knowledge never made it to long-term declarative memory, it was never truly learned at all (only memorized in the short term)."[14] The race to do this again and again often leads students to feel as if they are, what Denise Pope of Stanford University terms, "doing school," merely complying, as opposed to learning—a topic we discuss more in chapter 10, and which also can have implications for stress and health. Research in educational neuroscience, therefore, should not only inform the types of assessments teachers give students but also help shape

Table 9.1.   Assessment Tic-Tac-Toe

Podcast	3D Medium (trifold poster, diorama)	PowerPoint/Prezi/ Keynote Presentation
Student-Designed Flipped Classroom Lesson	Free Space (What else is available that your teacher has not learned about yet?)	iMovie/Final Cut
Public Speaking/Oral Presentation	Hypertext Essay	2D Medium (painting, drawing, PowerPoint poster)

Note: The assessment options above can be used with middle-school and high-school students. Assessment Tic-Tac-Toe is designed to offer students options that both play to their strengths and encourage the development of new skills. During the school year, each student must score tic-tac-toe by choosing a row, column, or diagonal assessment "chain" to be used as a summative assessment for three different units. Students are responsible for creating a grading rubric for each assessment they select as part of their assessment choice.

the strategies students use to prepare for assessments, or to work through projects, thus making these true learning opportunities.

How can all this research impact assessment at your school? Think of this as a challenge to teachers to embed richness in assessments—to think about how to foster optimal long-term learning when designing assessments for their classes or for individual students. We recognize that students—*all* students—have areas of evolving strength and weakness, passion and disinterest. When we hit the learning sweet spot, we both challenge and support all students. What is important to understand about the human brain is that it changes with experience, a process often referred to as plasticity. When we create rich assessments that require deep thinking and problem solving, teachers are expanding students' cognitive capacities. How exciting to think of teachers as "brain changers."

Every year, we need to assess every student in multiple, developmentally appropriate ways (see table 9.1). This is what differentiated assessment means. Some assessments will play to a student's strengths while others will pose significant challenges; some assessments will, wonderfully, do both.

Another central value of alternative assessments is that they help students learn essential skills for success in today's world—such as critical thinking, problem solving, communication, collaboration, resiliency, and grit. The critical value of these skills becomes apparent when we consider Tony Wagner's "Seven Survival Skills" to thrive at work in the twenty-first century.[15] Where do students hone, or even, in some cases, discover these skills?

- Critical thinking and problem solving
- Collaboration across networks and leading by influence

- Agility and adaptability
- Initiative and entrepreneurship
- Effective oral and written communication
- Accessing and analyzing information
- Curiosity and imagination

These skills are best developed through a variety of alternative assessments, particularly through project-based learning and design challenges in which "failing forward"—mistakes or failures that lead to breakthroughs, understanding, and student growth—is part of the assessment process.[16]

Well-crafted projects enhance student engagement. We also know that when students can own their learning choices, and make an emotional connection to the material, their learning is enhanced. Moreover, when we challenge students to demonstrate their learning in an authentic, purposeful way, they become more engaged in the process and actually learn more. As University of Virginia psychology professor Daniel Willingham points out, "A teacher's goal should always be to get students to think about meaning."[17] Alternative assessments do just that.

Let us also look at work from the University of Chicago Consortium on Chicago School Research (CCSR), who created a framework of noncognitive factors linked to a student's long-term success based on a large metastudy of existing research.[18] They found five general categories of noncognitive factors linked to success:

- Academic behaviors: attending class, being engaged, participating in class, completing assignments
- Academic perseverance: grit, tenacity, self-control, delayed gratification
- Social skills: interpersonal qualities such as cooperation, assertion, responsibility, and empathy
- Learning strategies: study skills, metacognition, self-regulation, goal-setting
- Academic mindsets: four academic mindsets are shown to contribute to academic performance: (1) *I belong in this academic community* (sense of belonging); (2) *My ability and competence grow with my effort* (implicit theories of ability); (3) *I can succeed at this* (self-efficacy); (4) *This work has value for me* (expectancy-value theory)[19]

While the standard palette of assessments found in most schools might address *some* of these, to equip our students with the greatest chance for success we should be assessing them in a broad variety of thoughtfully crafted ways that help develop *all* of these. Great projects again help here (and we must remember that great projects often evolve reflectively, iteratively, and collaboratively over time).

Furthermore, if we look again at the list, many of the principles of MBE research we have discussed in this book apply here too. So this is our challenge to you: as you plan the assessments you intend to use over the course of the year, how are you going to develop Wagner's Seven Survival Skills? How are you going to develop CCSR's five noncognitive factors? Look beyond your course—you are preparing your students to be successful for their long, interesting, uncertain journeys beyond your room.

There are other day-to-day things that can be done. Here are some important, research-informed changes in the ways teachers think about assessment for their students. Such translation of research to practice further extends a school's highest-achieving students and improves student achievement overall.

- Most units begin with a pretest to assess prior knowledge and to be able to measure individual student growth from the beginning to the end of a unit.
- Teachers use more formative assessments, such as ungraded (thus low-stress) surprise quizzes, as a self-reflective process for students. Frequent formative assessments, which can be as short as one question, allow students to practice recalling knowledge from their long- and short-term memory. Frequent retrieval of information significantly enhances recall ability. This is called the "testing effect." It is a great way to have students begin or exit a class. There are technology tools, such as socrative.com, designed specifically to help with this process.
- Teachers provide students with test-correction opportunities. Delaying or "scaffolding" feedback, and having students struggle with finding the correct answer, leads to better retention than does simply providing correct answers.
- In order to help students decide appropriate study strategies, teachers ask them to reflect on what brain demands—attention, memory, visual thinking, executive functioning, language, neuromotor function, social or higher-order cognition—a particular assessment might place on their brains.
- After an assessment, students are provided with an opportunity to reflect—think metacognitively—on their assessment performance, particularly on how well their study strategies worked and how they might

study differently in the future. Thinking about what brain demands were in fact used can add new depth to this reflection.
- Teachers shorten the distance and increase the frequency between when a concept is first introduced and then recalled and applied to new learning.
- Teachers share research into the transaction costs associated with multitasking and how limiting the toggling between websites or social media channels while studying can enhance memory consolidation (certainly no easy request of today's students).[20]

THE BENEFIT OF A NEURODEVELOPMENTAL FRAMEWORK

Our school benefited from having every teacher trained in a neurodevelopmental framework—for us it was the All Kinds of Mind (AKOM) neurodevelopmental framework for learning. But perhaps the main benefit to us was not one that AKOM intended. It led to a great amount of thoughtful differentiation of teaching and assessment.

AKOM presents a framework of brain demands with three hierarchical levels. The top level consists of eight constructs: Attention, Memory, Language, Temporal-Sequential Ordering, Spatial Ordering, Neuromotor Functions, Social Cognition, and Higher-Order Cognition. There are two deeper levels of terms, but we will not go into them here. The "entry level" goal of AKOM is to use this framework to analyze the strengths and weaknesses of each individual student, based on observational evidence. As the St. Andrew's faculty used the AKOM framework, however, it became clear that its real power came about when using it as a lens to examine, together, the following five factors:

- Each student's strengths
- Each student's weaknesses
- The brain demands germane to your subject
- The brain demands of how you teach
- The brain demands of how you assess

If you bring the last three of these into some sort of alignment, there are an authenticity and essential fairness that are not lost on the students—this alone is a powerful statement by a teacher.

But this lens also allows the teacher to vary assessments from week to week so that the same brain demands are not worked and reworked again and again, while others remain barely touched. In varying this, a different set of students get challenged and supported from one assignment to another. How do you challenge a student with strong receptive language and memory skills, the kind of student who remembers a great deal of what he or she hears and reads and for whom "traditional" school tasks are not super taxing (he or she reads the textbook, listens in class, and readily repeats information back on traditional tests)?

One way is to challenge other brain demands that this student may not *currently* be so strong in. "Currently" is a key word here—remember, the brains of our students have a significant degree of plasticity, and deliberate practice can rewire neurons to make them stronger learners in a broader range of brain demands.

Using the AKOM lens to view these five factors led to a large amount of differentiated instruction and assessment at St. Andrew's, and really drove a transformative discussion for the school: What is great teaching? And, indeed, what is learning?

As designers of their classes, teachers decide the appropriate assessments for a certain body of knowledge or skill. Using research from MBE science to inform how they do this should help them do it even better. But what research regarding choice and engagement shows is that teachers should also respect the ability of *students* to make decisions about how they can best demonstrate their understanding (as well as their confusion). Choice need not be part of every assessment, perhaps, but pick the moments. Could you imagine that happening in your class? That is why it is essential that teachers and educational leaders have ongoing professional development about how the brain learns, works, and changes.

Research on teacher efficacy points to the importance of expanding each teacher's toolkit with an MBE science lens into curriculum design and working with every student.[21] Such research supports the notion that when teachers understand the principles from educational neuroscience, they expand their teaching repertoire, including the ways they measure and differentiate how they measure a student's knowledge, skills, and understanding. Assessments today, therefore, are much broader than what most of us adults experienced in our own academic journeys. That is a good thing.

Without looking back from this page, what are the *three* most salient points you take away from this chapter of *Neuroteach*?

What are *two* things you would like to do "tomorrow" with the information you learned from reading this chapter?

What is *one* question you have after reading this chapter?

NOTES

Portions of this chapter originally appeared in a coauthored article by Glenn Whitman and Dr. Mariale Hardiman titled "Assessment and the Learning Brain," *Independent School Magazine* (Winter 2013). Dr. Hardiman is cofounder and director of the School of Education's Neuro-Education Initiative (NEI) at Johns Hopkins University and the author of *The Brain-Targeted Teaching Model for 21st-Century Schools*.

1. Daniel T. Willingham, *Why Don't Students Like School? A Cognitive Scientist Answers Questions About How the Mind Works and What It Means for the Classroom* (San Francisco: Jossey-Bass, 2009).
2. The final-exam idea was adapted from Jim Percoco's "Historical Head" project in *A Passion for the Past: Creative Teaching of US History* (Portsmouth, NH: Heinemann, 1998).
3. See Mariale Hardiman, *The Brain-Targeted Teaching Model for 21st-Century Schools* (Thousand Oaks, CA: Corwin, 2012).
4. Carole Ames and Jennifer Archer, "Achievement Goals in the Classroom: Students' Learning Strategies and Motivation Process," *Journal of Educational Psychology* 80, no. 3 (1988): 260–67.
5. Paul R. Pintrich, "Multiple Goals, Multiple Pathways: The Role of Goal Orientation in Learning and Achievement," *Journal of Educational Psychology* 92, no. 3 (2000): 544–55; Heidi Grant and Carol Dweck, "Clarifying Achievement Goals and Their Impact," *Journal of Personality and Social Psychology* 85, no. 3 (2003): 541–53.
6. Harold Pashler, Nicholas J. Cepeda, John T. Wixted, and Doug Rohrer, "When Does Feedback Facilitate Learning of Words?" *Journal of Experimental Psychology: Learning, Memory and Cognition* 31, no. 1 (2005): 3–8.
7. Keri L. Kettle and Gerald Häubl, "Motivation by Anticipation: Expecting Rapid Feedback Enhances Performance," *Psychological Science* 21, no. 4 (2010): 545–47.
8. Lisa K. Fazio, Barbie J. Huelser, Aaron Johnson, and Elizabeth J. Marsh, "Receiving Right/Wrong Feedback: Consequences for Learning," *Memory* 18, no. 3 (2010): 335–50.
9. See, for example, Nate Kornell, Alan D. Castel, Teal S. Eich, and Robert A. Bjork, "Spacing as the Friend of Both Memory and Induction in Young and Older Adults," *Psychology and Aging* 25, no. 2 (2010): 498–503.
10. See, for example, Jeffrey D. Karpicke and Henry L. Roediger, "The Critical Importance of Retrieval for Learning," *Science* 319, no. 5865 (2008): 966–68.
11. Jeffrey D. Karpicke and Janell R. Blunt, "Retrieval Practice Produces More Learning Than Elaborative Studying with Concept Mapping," *Science* (2011), doi:10.1126/science.1199327.
12. Nate Kornell and Lisa K. Son, "Learners' Choices and Beliefs about Self-Testing," *Memory* 17, no. 5 (2009): 493–501.
13. Sanjoy Mahajan, "To Develop Expertise, Motivation Is Necessary but Insufficient," Freakonomics.com, September 27, 2015, http://freakonomics.com/2011/11/25/to-develop-expertise-motivation-is-necessary-but-insufficient.
14. Tracey Tokuhama-Espinosa, *Mind, Brain, and Education Science: A Comprehensive Guide to the New Brain-Based Teaching* (New York: Norton, 2010).

15. Tony Wagner, *The Global Achievement Gap: Why Even Our Best Schools Don't Teach the New Survival Skills Our Children Need—And What We Can Do about It*, first trade paper edition (New York: Basic Books, 2010).

16. Paul Tough, *How Children Succeed: Grit, Curiosity, and the Hidden Power of Character* (New York: Houghton Mifflin Harcourt, 2013). See also Ken Robinson and Lou Aronica, *Creative Schools: The Grassroots Revolution That's Transforming Education* (New York: Viking, 2015); and Grant Lichtman, *#EdJourney: A Roadmap to the Future of Education* (San Francisco: Jossey-Bass, 2014).

17. Willingham, *Why Don't Students Like School?*, 61.

18. C. Farrington, M. Roderick, E. Allensworth, J. Nagaoka, T. Keyes, D. Johnson, and N. Beechum, *Teaching Adolescents to Become Learners: The Role of Noncognitive Factors in Shaping School Performance: A Critical Literature Review* (Chicago: University of Chicago Consortium on Chicago School Research, 2012).

19. Text taken from: Jenny Nagaoka, Camille A. Farrington, Melissa Roderick, Elaine Allensworth, Tasha Seneca Keyes, David W. Johnson, and Nicole O. Beechum, "Readiness for College: The Role of Noncognitive Factors and Context," *Voices in Urban Education* 38 (2013): 45–52.

20. Eyal Ophir, Clifford Nass, and Anthony D. Wagner, "Cognitive Control in Media Multitaskers," *Proceedings of the National Academy of Science* 106 (2009): 15583–7.

21. Mariale Hardiman, Luke Rinne, R. JohnBull, Emma Gregory, and Julia Yarmolinskaya (in review), "How Knowledge of Neuro- and Cognitive Sciences Influences Teaching Practices, Attitudes, and Efficacy Beliefs," manuscript submitted for publication; R. JohnBull, Mariale Hardiman, and Luke Rinne, "Professional Development Effects on Teacher Efficacy: Exploring How Knowledge of Neuro- and Cognitive Sciences Changes Beliefs and Practice" (paper presented at the American Educational Research Association conference, San Francisco, 2013).

10

HOMEWORK, SLEEP, AND THE LEARNING BRAIN

I was made for the library, not the classroom. The classroom was a jail of other people's interests. The library was open, unending, free.

—Ta-Nehisi Coates

One of the most watched TED talks of all time is Sir Ken Robinson's "How Schools Kill Creativity." This is in part because he is a fabulously engaging presenter, but it cannot be all. His message must resonate. We would hazard that one area where people look at their own or their children's schooling and go "YES!" is homework.

Homework is a topic that brings passionate responses, ranging from "a rigorous school *has* to have rigorous homework," to calls for no homework at all. Bizarrely, the rigor of a school is often judged by the hours of homework its faculty require their students to do, which is a bit like judging a dog by how long it is able to bark. Barking is important; don't get me wrong, I would hate to have a dog that didn't bark. But not all the barks my dog makes are the same, and some just seem better than others. Think back to your own schooling, think about the homework, all of it in all its many splendid forms. Then do the following three tasks:

(1) List three words that quickly come to mind when you think of homework.

(2) Write two questions that quickly come to mind when you think of homework.

(3) Write one metaphor that comes to mind when you think of homework.[1]

What made great homework for you? What made horrible homework? What would you have done with an evening off with no homework to do? Mention the word "homework" in a group of adolescents—just that one word—and carefully watch their faces. Homework fundamentally shapes the lives of many young people in part because it shapes the nature of their relationship with their family and friends, essential elements of being human, in the "free time" they have between their "job"—school—and sleep. In doing so, it fundamentally shapes the sleep they get, which has physiological, neurological, and psychological effects. Given this, we had better be really, really thoughtful about how we design programs of homework. We had better be using the best evidence possible to make decisions on how to do it well. Oops.[2]

ELEMENTARY SCHOOL

To make this discussion a little easier, we will consider elementary and secondary school separately. There has been a lot of study of homework in elementary school, but it has mostly focused on the correlation between homework and how well schools perform. These studies show that students in high-performing elementary schools tend to have homework, but they cannot say whether homework actually causes the higher performance or not.

There are far fewer studies that examine this causal relationship, and while many of these tend to show that homework is beneficial, "the evidence is less secure"[3] and the effect tends to be very small—according to Dr. Harris Cooper, Professor of Education and Hugo L. Blomquist, Professor of Psychology and Neuroscience at Duke University, "the average correlation between time spent on homework and achievement . . . hovered around zero."[4]

Part of the difficulty in honing in on the importance of homework is because "effective homework is associated with greater parental involvement and support,"[5] which ties homework in to other socioeconomic factors that might affect performance. It also speaks to the difficulty in crafting a study where the only factor you are studying is the effect of homework.

So what should homework in elementary school look like? Homework might be best "used as a short and focused intervention [where] it can be effective in improving students' attainment."[6] However, if homework is more routinely set, the benefits are likely to be "modest."[7] That is, homework should perhaps not be the rule, the daily grind, but assigned with great thought on occasion. Dr. Harris Cooper adds, "A little amount of homework may help elementary school students build study habits,"[8] and it may help students learn self-discipline and time-management skills[9]—habits and skills to help them be successful in later grades. But even this claim, which many people may feel makes intuitive sense, has scant evidence to support it.

Given the importance of factors like play[10] and reading for pleasure,[11] and the lack of overwhelming research evidence to support a rigorous regime of elementary-school homework, it seems prudent to err on the side of "little" and "not every day," and to cry out for more research to help move thinking on this important issue. We must, however, make a stand for what makes a "rigorous" elementary school, and suggest that it be one that values and holistically balances the social, emotional, and educational development of children in the crucially important elementary school years, prime years for certain brain developments, not one that sets hours of homework.

SECONDARY SCHOOL

The research on the effectiveness of homework in secondary schools is stronger. However, as the Education Endowment Foundation (EEF) reports, "beneath this average there is a wide variation in potential impact, suggesting that how homework is set is likely to be very important."[12] So, evidence suggests there are good ways and bad ways to do homework. Let's begin by thinking about a typical school homework regime.

There is a horrible irony with typical homework. High-flier kids might find it boring because they "get it" quickly so that it feels like busy work, and "downshifting" probably kicks in to some degree, with the amygdala flagging it as not worthy of serious prefrontal cortex higher-order-thinking processing power. To "challenge" such students, teachers often retaliate with quantity rather than quality, and miss the point that the additional work is probably seen as just as tedious as the original. Significant, challenging, worthy ideas typically require more than twenty minutes, but homework, done well, can play a role in teeing up the further thinking that needs to be done. Done badly, it does not only fail to tee this up, it also fills all possible "mulling-over time" with vacuous noise.

Conversely, the students who are more challenged by learning may often find themselves feeling overwhelmed. Picture a student who is taking a regular load of six or seven classes. In the typical daily schedule this student will have six or seven lots of homework to do. Let us be generous and say that each teacher sets thirty-minute assignments, but one teacher may set nothing, which makes three hours of homework.

Now, given how many tests a student typically has to take, thirty minutes for every subject may not be realistic, but let us just take the generous scenario and say that it is, so it still makes three hours of homework. Then add this on top of the day the student has just had—six or seven different subjects that he or she has been rushing between, each with different brain demands and different stores of long-term memory with which he or she needed to interact.

Then add in some extracurriculars: a sport practice, dance class, music lesson, maybe more than one commitment; there needs to be room for something in students' lives beyond the classroom that brings them joy, that supports a passion. Somewhere along the line, the student grabs food, checks in with family, connects with friends—all these are important things that are deeply set human needs. Finally, the student needs to somehow turn his or her frantically spinning mind off and sleep. Then, all too soon, the student must wake and repeat. It has to be exhausting.

We tried shadowing students, and suggest you do, too. Even for just for one day, even without the extracurriculars, just get to the end of that day, walk in through your front door, check how tired you are feeling, then look at all the work you still are expected to do, and will be penalized and punished if you do not. And imagine that this continues day after day, and add in other anxiety triggers of adolescence, like relationships with peers and

Figure 10.1. Balancing sleep with other demands, including those of school, is often hard, even when both students and teachers are aware of the important role sleep plays in learning.

family, forming and finding one's own identity, or contemplating the next step of life when high school is done.

Now picture what a student with an extended-time accommodation—maybe because of slow processing, reading or written elaboration speed—is facing. How long will it take this student to do homework? If he or she were to do it all well, four to six hours maybe? Think about all the things you faced when you shadowed a student—they are likely to be that step harder. Now imagine how tired this student will be at the end of the day as they sit down to begin the homework pile.

One horrible irony, therefore, is that the typical regime of homework often leaves all types of student, from the high flier to the student facing learning challenges, feeling annoyed. It is the great unifier of education.

Another horrible irony is that the activities students really need to do to be successful, like check in with a teacher or peer tutor, have enough time to do spaced active retrieval over a period of many days to help memorize material, or have adequate sleep (which research shows is crucial for memory consolidation), tend to be those that students don't do or stop doing when they feel so rushed because of their homework load. They are so busy meeting the "get it done" requirement set by their teacher that the thought of adding one more jot of work to their pile, even if it might lead to increased learning, is often too much to contemplate.

Some students set the bar low to survive; they are not aiming for learning, but rather getting by so that their grade does not suffer. To illustrate this, these are two comments that have not been carefully picked from hours of interviews, but rather have been picked because they are things that we happened to hear yesterday: "I'm not even thinking when I'm doing it, I'm just trying to get it all done." "There's no way I'm trying for mastery, I just don't want to get a zero." Listening to high-school students talk about homework is enlightening—especially when you think to yourself, "Where are they setting the bar for learning? How does learning fit in with school?" Or even, "What is the fundamental purpose of school?"

In the face of this, some students just give up. Many have high rates of anxiety. We must question, at what point, or in how many cases, does stress exacerbated by homework contribute to "toxic stress," characterized by consistent high levels of stress where the body's stress-response system does not get a chance to reset, which we know has physiological effects with long-term, even lifelong, negative health consequences?

There is a great, multifaceted bell curve of student abilities, so multidimensional that it could not be drawn on a piece of paper. Which students in this rich spectrum have typical homework regimes really been designed to serve? We suggest it is all too skinny a fraction. We can do better.

EEF's analysis of the research literature on homework in secondary school makes the following observations.[13]

- "There is some evidence that homework is most effective when used as a short and focused intervention (e.g., in the form of a project or specific target connected with a particular element of learning) with some exceptional studies showing up to eight months' positive impact on attainment."
- "Benefits are likely to be more modest, up to two to three months' progress on average, if homework is more routinely set (e.g., learning vocabulary or completing problem sheets in mathematics every day)."
- "Homework is most effective when it is "an integral part of learning, rather than an add-on."
- Providing high-quality and timely feedback on homework is important.
- "There is an optimum amount of homework of between one and two hours per school day (slightly longer for older pupils), with effects diminishing as the time that students spend on homework increases."

- "[Homework] should not be used as a punishment or penalty for poor performance."
- "The quality of homework is more important than the quantity."
- "A variety of tasks with different levels of challenge is likely to be beneficial."
- Making the purpose of the homework assignment clear to students is beneficial "(e.g., to increase a specific area of knowledge, or fluency in a particular area)."

EEF also gives the important caveat that most studies on the relationship between homework and achievement discover just a correlation, not causal evidence, which is much harder to come by. "It is certainly the case that schools whose pupils do homework tend to perform well, but it is less clear that the homework is the reason why they are successful."

MBE RESEARCH

In addition to this, what ideas from MBE research can also inform the great homework debate? This is an opportune time to reflect back on the previous chapters of this book, and so we are going to give you a chance to do just that.

What ideas from MBE research can help us make homework better?

We suggest some familiar MBE themes are relevant here—for example, stress, sleep, choice, novelty, teaching and assessing in multiple modalities, using modalities that are driven by content, using arts integration, requiring and helping structure executive functioning tasks, the use of play, giving rapid feedback, giving scaffolded feedback, creating a safe place to get something wrong, the effect of emotions, the importance of relationships, and the need for hard work and smart work to have the greatest positive influence on neuroplasticity.

1. Remember that boredom has an effect on the amygdala that sends incoming sensory information to the "flight, flight, freeze" reactive

part of the brain rather than the reflective part of the brain where higher order thinking and executive functioning take place. Remember the Yerkes-Dodson curve—that a certain degree of "arousal" is required to increase attention and interest, but too much impairs attention, working memory, and decision making. In between, the sweet spot of optimal performance, is something we called "the zone of proximal discomfort."

2. Remember that a certain amount of tolerable stress is a good thing as it helps developing bodies and minds build robust stress systems to deal with future stresses—but the important points are that students have supportive relationships, and that the stress is episodic rather than never ending.

3. Intrinsic motivation and engagement improve learning, and while these factors can sometimes elude our best efforts, chances of fostering them may be improved by strategies like giving students choice, adding novelty, adding relevancy to their lives, incorporating well-chosen aspects of play, including things that lead to students making an emotional connection, and creating a school culture that stresses the importance of positive peer–peer and peer–teacher relationships.

4. The primacy-recency effect suggests that students may remember most what they hear at the beginning of class—so do not religiously use this precious time to go over last night's homework. At times you may want to emphasize a point, but make this the exception not the rule so that its novelty stands out.

5. Rapid feedback aids motivation and engagement and leads to improved learning, as does scaffolded feedback and the opportunity to redo work. This suggests that homework that is done, merely checked off that it is done, and then is not heard of ever again, is not great homework. Getting some feedback is key, even it is something as simple as giving access to worked answers. Assignments that are returned by teachers weeks and weeks after being turned in, even if they are full of excellent feedback in whatever color of pen, do not result in great learning. In fact, research shows that even the anticipation of rapid feedback leads to better learning.

6. Sleep is crucial. Lack of sleep deteriorates a wide swathe of brain performance, including working memory function, long-term memory storage, and memory retrieval. Whether we like it or not, teachers play a role here by how much homework we assign and how we schedule it.

7. Teachers tend to believe that in order to be a rigorous teacher they have to assign homework every night; this is a fallacy. Quality of

homework is more important than quantity. Expert teachers assign great homework, not lots of homework—just the right task at just the right time. It is the arthroscopic-keyhole-knee-surgery approach to homework, as opposed to the "let's just open it all up and have a rummage around" one.

Research suggests that one to two hours of homework a night is optimal in secondary school—maybe up to two and a half hours for the oldest students. Beyond this the effect of homework actually tends to diminish. So if a student has six or seven teachers, does each assign only fifteen to twenty minutes? Or do some not set homework? Or do you create some type of block schedule that reduces the number of classes for which a student has to prepare each day? It may sound extreme to change schedules based on homework, but it may sound less extreme if you take the time to listen to students talk honestly and thoughtfully about learning.

8. To aid the goal of sleep, teachers have a role in helping students develop a system that works for them that makes sure assignments are both completed and turned in. Yes, parents do and should play a huge role here, but as we will discuss below, deliberate work to grow executive functioning skills tends to be effort well spent.

9. Remember that our students have at least fifteen more years of significant brain development in their prefrontal cortex, the area responsible for executive functioning, so include scaffolding to help them develop the skills of organizing, planning, executing, evaluating progress, and adjusting accordingly. Remove this scaffolding as appropriate, realizing that different students need different scaffolding, with the goal of putting each student in their zone of proximal discomfort.

    This is effort well spent. One day, even if it as far ahead as the third year of med school or law school, they may thank us. Elementary schools tend to be really good at deliberately teaching these skills, but secondary schools tend to ignore them more, mistakenly thinking that these skills are already well developed. However well developed they are, they have a good many years of prime neuroplasticity to make them even better.

10. There is no such thing as multitasking. The brain cannot multitask—instead it rapidly switches from meeting the demands of one task to meeting the demands of another. There is a switching cost for doing so. Students (and adults) will insist that this is not true and that they can multitask—they are wrong. It results in more effort for less-efficient performance. We need to repeatedly coach students, and

parents where possible, to create homework environments where multitasking is limited.

11. Remember the "multiple intelligences" neuromyth. While each student has individual differences in brain demands that he or she is good at and not good at, it is a myth to try to teach to each individual student's strengths. Instead, the best modality to choose for teaching and assessment should be driven by content. Use this to help craft quality assignments.

    This intersects somewhat with the idea of using arts integration to increase learning. In both cases, be careful of whose work you are assessing, the parents' or the students', and design projects accordingly, even if it means "sacrificing" class time to bring some construction activities into the classroom. The skills you bring into the classroom by doing so, including things like collaboration, communication, and creativity, are valid brain demands that we as teachers should be actively fostering anyway.

    Done well, students will see that designing and building is a very social activity, and relationships are important in a community of problem solvers, which can help more people see activities like this as something that they can, and want, to do.

12. Memory tasks play a large role in secondary school, and there is nothing wrong with that. But we can help students by building in interleaving as we plan our weeks and months of homework, and encourage them to space their studying rather than use massed studying. We can also assign and promote active-retrieval approaches to studying, rather than reading and reading notes or textbooks, a practice that tends to lead to the illusion of fluency. Teaching or suggesting memorizing methods alongside material that needs to be memorized can be effective in modifying students' practice because the advice is timely and comes in context. One important strategy to remember to promote for memorization is the use of brain breaks and movement to reset attention.

13. Low-stakes or no-stakes formative assessments can be a good tool to assess what future homework could be (as well as working to improve memory storage themselves). This gives the students a feeling that homework assignments might be tailored to meet what they need, which can increase buy-in. Formative assessments also add variation to more routine practice assignments.

14. The act of getting something wrong is a key part of learning, prompting rewiring of the brain as students work to get it right. There are

several implications to this nugget of neuroscience. Homework is a great place to work on things at increasingly difficult levels until you get to the point where you get something wrong. Students need space to struggle with getting something right without being penalized for reaching the point where great learning is taking place, which suggests that assigning a grade based on demonstrating significant effort rather than on percent correct might be fairer if the goal is ultimately learning the subject. It also suggests, as we have stated above, the need for timely, scaffolded feedback and the opportunity to redo, if we assume that the role of homework is to aid learning.

15. What is working for a student with regard to homework routine? What is not? What can the student do about it? Reflection and meta-cognition activities as homework assignments may help shift practices and mindset. The amount and quality of effort are important to success, as is getting students to realize that they have the ability to rewire their brains to become better learners and higher achieving students by working hard and working smart.

    We will not get students to have a growth mindset by telling them "have a growth mindset!" no matter how many posters saying so we put up. But we have a chance of doing so by engaging them in a conversation on the amount and quality of their effort, and how it meets the demands of the subject. Homework plays an important role in this conversation.

16. Do you actively work to ensure that your students feel heard, listened to, and known? Learning and emotion are linked. When we are teaching well, we are pushing students into areas of thinking and levels or amounts of work where they do not feel comfortable. Their relationship with and trust in their teacher are a significant part of whether they will succeed—and thus a significant part of how far we can push each individual student in their own distinct "zone of proximal discomfort."

17. Relationships are really important in learning. When a student's perceived bulk-of-my-work focus shifts away from the classroom to homework, it shifts away from a place where relationships are central to learning to one where they are not.

As we have seen before, MBE science does not give a recipe book for a potions lesson at Hogwarts, and it is not a "do exactly this and you will get great homework" list. The context of a particular class, an individual teacher's pedagogical content knowledge, as well as his or her own unique voice as a teacher, are all factors that affect how this MBE research should be used to inform a teacher in their craft to create great homework. Again we get to this crucial point that the place of MBE research is to *inform* practice—the word "inform" leaves room for the craft of teaching. And teachers must always keep in mind that crafting great homework is an iterative practice—try, reflect, tweak, try again.

Good guiding words for homework come from Maurice Elias, Professor of Psychology and Director of Rutgers' Social-Emotional Learning Lab: "The real question we should be asking is, 'What do we believe should happen after the end of the school day to help ensure that students retain what they have learned and are primed to learn more?'"[14]

At first glance this seems to highlight the more typical day-to-day homework activities, urging us to do them effectively and efficiently. But it must also leave room for those well-chosen occasions when we want students to experience that magical feeling when they become so engrossed in an assignment that they lose track of time, or their stresses over workload fade away as they happily spend hours deeply engaged in a project (and later say that "it cannot have been *that* long" or "it did not really feel like work"). We bet that if we all look back over our own schooling, we can pick out times like these.

Psychologist Mihaly Csikszentmihalyi might describe such students as being in a state of "flow," a state of focused engagement with a highly challenging, highly skilled activity so intense it involves a temporary loss of self-awareness. Frequent experience of flow is linked to achievement, work satisfaction, and creativity. Katherine R. Von Culin, Eli Tsukayama, and Angela L. Duckworth suggest that people motivated by flow are more likely to seek activities that challenge their skills and abilities, and more likely to maintain effort toward long-term goals. This links creating opportunities for flow with helping develop what Angela Duckworth calls "grit."[15]

When people are driven by engagement in an activity and flow, it seems to facilitate sustained effort over time, a major factor of grit. When people are driven by immediate pleasure, however, it seems to undermine sustained, focused interest over time, decreasing grit. It suggests a need to carefully craft such opportunities for students—which ties in with homework, the types of work we value, and the projects we craft.

"Activities that involve authentic challenge, self-expression, interpersonal relationships, problem solving or competitive sport appear to be particularly effective in promoting flow experiences."[16] As well as deep levels of engagement, intrinsic motivation, and sustained effort, flow is linked to happiness as a student. There is also emerging research to suggest that happiness is related to achievement, and that this relationship is probably causal in both directions.

We are not saying that all homework should be flow-inducing like this, far from it, but we must make sure these moments exist, and exist in the right spots. Our list of MBE strategies above gives hints on how to distinguish them from "just another boring, time-consuming project," but this is also an area where subject knowledge and pedagogical content knowledge help mightily.

Is all this a call to make homework easier, more trivial? To answer, think of the story of Sisyphus, which may be seen as a metaphor for more traditional approaches to homework. Yes, "expert" homework may be easier, if you mean that it reduces the Sisyphus mindset, where students feel doomed to spend eternity pushing a boulder of worksheets, book reports, tri-folds, and test study guides up a hill with a scuffed "academic rigor" signpost wonkily sticking out of it. But it does so precisely because it is anything but trivial; it is the opposite of the Sisyphus approach to homework. It represents a great mindfulness toward how we best extend a student's day beyond the classroom.

Furthermore, on any given day, in any given context, it may be the best decision to set no homework at all. The expert teacher is willing and has the

Figure 10.2. The Sisyphus mindset.

judgment to make that call. The expert teacher assigns great homework, not lots of homework; they define themselves by the quality, not the amount. The rigorous school has expert teachers, not teachers who compensate for their lack of expertise by assigning the biggest pile of homework in the town.

Denise Pope, senior lecturer at Stanford's Graduate School of Education, highlights a problem that she calls "doing school":

> These students explain that they are busy at what they call "doing school." They realize that they are caught in a system where achievement depends more on "doing"—going through the correct motions—than on learning and engaging with the curriculum. Instead of thinking deeply about the content of their courses and delving into projects and assignments, the students focus on managing the workload and honing strategies that will help them to achieve high grades.[17]

Think about the role that homework plays in this. By our very nature, by our evolutionary history, our minds are intended to learn. So it should not surprise us that when learning happens, it is fun. Learning is the antithesis of "doing school." We need to refocus ourselves to be in the learning business, not the "doing school" business.

We leave you by sharing this excerpt from a newspaper article written by Carolyn Walworth, a high-school junior and a student representative on her school district's Board of Education, which, we suspect, eloquently expresses the feeling of many.

> Telling us to go see a school counselor for stress is insufficient. It is analogous to putting a BandAid over a fresh gunshot wound. Students in our district understand how to cope with stress; the real problem is that they simply have too much of it to cope with. Students are gasping for air, lacking the time to draw a measly breath in . . . We are not teenagers. We are lifeless bodies in a system that breeds competition, hatred, and discourages teamwork and genuine learning. We lack sincere passion. We are sick. We, as a community, have completely lost sight of what it means to learn and receive an education . . .
>
> It is time to rethink the way we teach students. It is time to reevaluate and enforce our homework policy. It is time to impose harsher punishments upon teachers who do not comply with district standards such as not assigning homework during finals review time. It is time we wake up to the reality that . . . students teeter on the verge of mental exhaustion every single day. It is time to realize that we work our students to death. It is time to hold school officials accountable. Right now is the time to act.
>
> Effective education does not have to correlate to more stress. Taking an advanced course should not be synonymous with copious amounts of homework.

Challenging oneself academically and intellectually should be about just that—a mental challenge which involves understanding concepts at a deeper level. The ever increasing intertwinement between advanced courses and excessive homework baffles me; indeed, I would say that it only demonstrates our district's shortcomings, our teachers' inabilities to teach complex materials in a way that students are entertained by and can understand. Instead, they rely on excessive homework to do the teaching for them.

These are issues that absolutely cannot wait. Please, no more endless discussions about what exactly it is that is wrong with our schools, and, above all, no more empty promises. Students live under constant stress every day of their lives. It is time to get to work.[18]

As people who care about the craft and profession of teaching, we cry out with impassioned voices, let's begin better serving Carolyn and all the countless ranks of students behind her.

Without looking back from this page, what are the *three* most salient points you take away from this chapter of *Neuroteach*?

What are *two* things you would like to do "tomorrow" with the information you learned from reading this chapter?

What is *one* question you have after reading this chapter?

NOTES

1. A "3 2 1 bridge" exercise from Dr. Ron Ritchhart's "Visible Thinking." Ron Ritch-hart, Mark Church, and Karin Morrison, *Making Thinking Visible: How to Promote Engagement, Understanding, and Independence for All Learners*, first edition (San Francisco: Jossey-Bass, 2011).

2. To find good evidence on homework, one starting point is The Education En-dowment Foundation (EEF, www.educationendowmentfoundation.org.uk), a nonprofit organization in the United Kingdom "dedicated to breaking the link between family in-come and educational achievement." It works to identify, evaluate supporting evidence, and encourage the adoption of effective educational practices that improve learning. Its "Teaching and Learning Toolkit" is a very accessible entrance point to the academic research surrounding many teaching practices, and thoroughly worth spending some time reading through. Its summary of the effects of homework is particularly useful.

3. "Homework (Primary) | Teaching and Learning Toolkit | The Education Endow-ment Foundation," https://educationendowmentfoundation.org.uk/toolkit/toolkit-a-z/ homework/ (accessed May 13, 2015).

4. Harris M. Cooper, *The Battle over Homework: Common Ground for Administra-tors, Teachers, and Parents*, third edition (Thousand Oaks, CA: Corwin, 2006).

5. "Homework (Primary) | Teaching and Learning Toolkit | The Education Endow-ment Foundation."

6. "Homework (Primary) | Teaching and Learning Toolkit | The Education Endow-ment Foundation."

7. "Homework (Primary) | Teaching and Learning Toolkit | The Education Endow-ment Foundation."

8. "Does Homework Improve Academic Achievement? If So, How Much Is Best?" *SEDL Letter* 20, no. 2, August 2008, http://www.sedl.org/pubs/sedl-letter/v20n02/ homework.html (accessed May 13, 2015).

9. Lyn Corno and Xu Jianzhong, "Homework as the Job of Childhood," *Theory into Practice* 43, no. 3 (2004): 227–33.

10. Stuart Brown, *Play: How It Shapes the Brain, Opens the Imagination, and Invigorates the Soul* (New York: Avery Trade, 2010); Daniel Willingham, *Why Don't Students Like School? A Cognitive Scientist Answers Questions about How the Mind Works and What It Means for the Classroom* (San Francisco: Jossey-Bass, 2009); L. S. Vygotsky, *Mind in Society: The Development of Higher Mental Processes* (Cambridge, MA: Harvard University Press, 1978).

11. Christina Clark and Kate Rumbold, "Reading for Pleasure: A Research Over-view," National Literacy Trust, 2006, *Google Scholar*.

12. "Homework (Primary) | Teaching and Learning Toolkit | The Education Endow-ment Foundation."

13. "Homework (Secondary) | Teaching and Learning Toolkit | The Education En-dowment Foundation."

14. http://www.edutopia.org/blog/homework-vs-no-homework-wrong-question -maurice-elias?webSyncID=581000c2-be69-e1c6-632e-253c0cae1565&sessionGUID =e1ea3006-e574-e150-de84-651dc37de059.

15. Katherine R. Von Culin, Eli Tsukayama, and Angela L. Duckworth, "Unpacking Grit: Motivational Correlates of Perseverance and Passion for Long-Term Goals," *The Journal of Positive Psychology* 9, no. 4 (2014): 306–12.

16. "Why We Must Have Joyful and Imaginative Learners | Considered—The Faculty of Education Blog at Canterbury Christ Church University," http://www.consider-ed .org.uk/why-we-must-have-joyful-and-imaginative-learners/ (accessed May 19, 2015).

17. Denise Clark Pope, *Doing School: How We Are Creating a Generation of Stressed-Out, Materialistic, and Miseducated Students* (New Haven, CT: Yale University Press, 2003).

18. "Paly School Board Rep: 'The Sorrows of Young Palo Altans,'" http://www.palo altoonline.com/news/2015/03/25/guest-opinion-the-sorrows-of-young-palo-altans (accessed May 21, 2015).

TECHNOLOGY AND A
STUDENT'S SECOND BRAIN

Midway through the exam, Allen pulls out a bigger brain.

—Gary Larson, caption to a *Far Side* cartoon
(We will leave you to go find it!)

Is it possible that two brains are better than one? Maybe yes, maybe no. For today's students, the ubiquity of electronic devices allows them to create, connect, and collaborate in ways beyond the imagination to many of their teachers. That is exciting news, but all schools have to wrestle with the challenge of determining the proper amount of technology, and when technology is helping learning or merely a distraction from it.

Dr. Seymour Papert is the long-time professor of learning research at MIT, cofounder of MIT's Artificial Intelligence lab and hugely influential Media Lab, and a visionary on how technology can influence pedagogy for children. He is considered by many to be the "father of educational computing." It is fitting that he opens this chapter:

Across the globe there is a love affair between children and the digital technologies. They love the computers, they love the phones, they love the game machines, and—most relevantly here—their love translates into a willingness to do a prodigious quantity of learning. The idea that this love might be mobilized in the service of the goals of educators has escaped no one. Unfortunately, it is so tempting that great energy and money has been

Figure 11.1. A student's second brain?

poured into doing it in superficial and self-defeating ways—such as trying to trick children into learning what they have rejected by embedding it in a game. Nobody is fooled. The goal should not be to sugar coat the math they hate but offer them a math they can love.[1]

It is a quote that hints at the potential for technology to captivate students and enhance learning. But it warns us that technology's easy ability to create an alluring learning package (and if you have spent any time at all with educational technologies you will no doubt be able to rapidly fill your mind with gorgeously crafted images and, probably, snippets of catchy tunes) is an insufficient goal. Instead, take "offer them a math they can love" as a metaphor for all subjects, all ages, all abilities that speaks of technology's potential to truly transform teaching and learning. Here is Professor Papert again:

So, too, the mega-change in education that will undoubtedly come in the next few decades will not be a "reform" in the sense of a deliberate attempt to impose a new designed structure. My confidence in making this statement is based on two factors: (1) forces are at work that put the old structure in increasing dissonance with the society of which it is ultimately a part, and (2) ideas and technologies needed to build new structures are becoming increasingly available . . . Public discussion of the idea-averse nature of School makes the dissonance more acute. Public access to empowered forms of ideas and the ways in which technology can support them fertilizes the process of new growth.[2]

Technology is a potential agent for "mega-change" in education, not because it is or should be the focus of the change itself, but because it is the catalyst for a change in our ability to access, manipulate, and communicate ideas. The drive for this change in education is not that kids' lives are full of technological marvels; it is that the way they interact with ideas in their everyday world has shifted to create dissonance with the way schools typically have them interact with ideas.

The same is also true for how they interact with people. So putting a gleaming new iProduct in every student's hands is not going to transform education by itself; what is needed is a pedagogical rethinking that acknowledges and utilizes what can now be achieved with technology as a partner in our teaching repertoire.

Two big red flags immediately come to mind. First, technology is hard and fast. Hard, as in immensely technical and costly to create, so that teachers will primarily be the consumers rather than the creators of new technologies. Fast, as in changing rapidly so that it is hard to fathom what will come next. For example, many people are familiar with Moore's Law, which predicts rapidly increasing computer-processing power. Another way to think of this, though, is that "sufficient" processing power will become ever cheaper (and smaller, and less power-consuming) so that it will appear in an increasing number of new, and often hard-to-predict, places.

Less well known, but at least as important to education, is Kryder's Law, which is a similar concept to Moore's Law but predicts the rapid increase in data storage capacity. In other words, our ability to store more and more stuff more and more cheaply in a smaller and smaller physical space. For example, within our lifetime we can predict an iPhone-like device, in both cost and size, that will be able to record and store a nonstop video recording of a person's entire life. And this ability already exists "in the cloud."

"Hard and fast" means that in the course of their careers, teachers will be consumers of technologies that they mostly cannot predict right now. It is hard to remain sane in what seems like it will be an overwhelming and never-ending cavalcade of technology and obsolescence. This book, however, is not intended to be a treatise on technology and education, it is not a "use this tool to do this . . ." guide. So, at the risk of seeming like we are dodging a charging hippo, we will put this strand aside with a cautionary note: "prepare to be agile."

Second, technology can be all-consuming. Perhaps the biggest message of our chapter on technology is that there are many times where it does not improve learning and so shouldn't be used. It may look great for a tour going past your classroom door, but remember, "sugar coating" is not our goal here. So our next cautionary note is: "be prudent and discerning." Our goal in this chapter is to use research from mind, brain, and education (MBE) to help us be so.

Daisy Christodoulou is a former teacher in the United Kingdom who followed her passion in investigating how research from cognitive psychology could inform her teaching and improve the learning of her students. Her first book, *Seven Myths about Education*,[3] may be little-known in the United States, but is a good read. Central to it is the role of knowledge in teaching, and how we can do a better job of handling knowledge. It is an assertive argument for a shift in pendulum swing to the contemporary "Why do I need to know it, I can just Google it?" mindset.

A base of knowledge is important at the outset of a task if students are to engage in higher-order thinking. Teachers have an important role to systematically teach knowledge—making sure that they are not making false assumptions about key little bits of knowledge, any one of which may prove to be a complete roadblock to a student's progression of understanding. Lecturing is one way to do this.

Yes, lecturing is okay, it has its place. Not all the time, but it should be part of a teacher's balanced repertoire (not least because it is something that students will most certainly encounter in college, so building a skillset before then to be as good with lectures as they can possibly be is a good thing). Technology can help in equipping students with a robust knowledge base, as a research tool or for various means of self-testing for example, but the main focus is on more traditional means of teaching to develop this knowledge base.

To help us think about the best places to use technology, we offer you a wonderfully elegant structure—one with ancient roots, but which has been revitalized by Martin Robinson, a teacher and school leader for more than twenty years, in his book *Trivium 21c: Preparing Young People for the Future with Lessons from the Past*.[4] "Trivium" is a phrase coined in the Middle Ages for the educational concept of grammar, dialectic, and rhetoric developed in ancient Greece. For a long time a cornerstone of a "classical" Western education, the essential nature of the trivium, as represented by Robinson, holds up well to MBE research. Robinson presents the trivium as a three-stage process for teachers to follow:

1. *Grammar*—learn the knowledge base.
2. *Dialectic*—use the knowledge base in discussion. (Robinson uses the word "discussion" in a broad sense to mean using, working, and manipulating the knowledge in some kind of participatory forum, not in the strictly narrow sense of just sitting around talking.)
3. *Rhetoric*—communicate the results of the discussion. (Again, Robinson uses "communicate" in a broad sense.)

As with Christodoulou's work, it starts by the teacher, purposefully and with structure, creating a base of knowledge (grammar). Then, following sound MBE research, it has the students use this knowledge in ways where they get feedback, which helps with a number of factors, including factual accuracy, engagement, and memory storage (dialectic). The broad use of the concept of "discussion" allows for bringing a broad range of modalities into the teaching, choosing the modality to best suit the particular content being studied.

Reworking the knowledge to communicate it (rhetoric) also follows sound MBE research: higher-order thinking is likely to be enhanced by students having a more deliberately created solid base to work from. Communication, too, allows for a diverse range of modalities for teaching and assessing to be used over the course of the year—again, with the teacher carefully choosing methods based on the content rather than perceptions of individual learning strengths or weaknesses, as well as with a thought to methods authentically germane to the discipline he or she is teaching. There is also the possibility for arts integration, which research suggests enhances engagement and memory storage. It is also easy to see how choice, novelty, and personal relevancy could, depending on the topic, be incorporated.

There is more, but we just want to give a reassuring taste that Robinson's work on the trivium is supported by MBE research, for two reasons. First, it is simple structure—easy to fix in your memory, easy to adapt to all kinds of situations. Second, it gives us a good structure for where we might best use technology. If the first stage, grammar, the knowledge base, is based on more "traditional" teaching methods supported by technology, the next two stages, dialectic and rhetoric, are where technology can really transform learning. It can do so by taking things that students currently do in other non-technology ways and allow them to do it better, and also by allowing them to do new things that they couldn't do before.

Think of the former as word processing an essay rather than hand writing it—word processing hasn't and should never replace hand writing entirely, but in its ability to rapidly and easily edit and re-edit work it has a place in improving the essay writing process. Think of the latter as Skype'ing a doctor working in Haiti's remote Central Plateau, interviewing him as part of a class design challenge to create something that will improve the quality of life there—possibilities that technology has opened up that, until we hear of others doing them, we often have a hard time conceiving them ourselves.

Let's look at the three stages of the trivium, apply an MBE research lens, and see where technology might enhance learning. In each, instead of discussing specific websites and applications, which will be superseded, we will focus on functionalities, which are still likely to be superseded but at a slower rate.

In grammar, technology can be used to help the core knowledge stick better in each student's memory. On the back side of content delivery, there are many applications to help students with active retrieval or self-testing. Rapid feedback should aid memory storage, as should scaffolded feedback that gives just enough nuggets of information and structure to the student and has them try to correct their mistake rather than just telling them the right answer. Many use novelty—just enough of Papert's "sugar coating" helps keep students in the game. The use of carefully constructed rewards systems is also something that MBE research suggests can increase attention and motivation.

Research suggests that rereading the textbook or class notes is a terrible way to study for an assessment as it tends to create an "illusion of fluency" rather than robust long-term memory storage—no matter how many highlighters a student uses. Technology has great potential to convince more students to rely more on active retrieval for studying.

On the front side of grammar, technology and its incredible access to vast vats of knowledge can allow for students' choice and self-agency—the feeling that they are in control of their own personal learning, which MBE research suggests might aid motivation. But the key to "grammar" is that they have the solid knowledge base necessary for dialectic. So the teacher has a key role to play in providing necessary scaffolding so that students efficiently learn all the things they need to learn—in effect, creating "the illusion of free choice."

The second stage of the trivium, dialectic, is discussion, but in its broadest sense. It involves students using and playing with ideas. It involves thinking, and thinking is good. Online discussion formats allow teachers to extend the school day, and the voices that push the discussion are often different—people who are quiet in class may feel more comfortable talking online. But doing this takes care and skill on the part of the teacher—the right amount and type of monitoring and feedback are crucial; students suffering from blog fatigue is a constant danger.

Shared documents can allow for collaborative back and forth and exist in many forms beyond a simple word-processing page that more readily invite the inclusion of different media. The role of the teacher is to make sure that these create discussion of the knowledge.

One example we want to share comes from Morgan Evans, an English teacher at St. Andrew's. He constantly wants his students to have an interesting thought that is worthy of deeper probing rather than going for something superficial or predictable. The problem is, essentially, how do you get a class of students to wrestle with ideas for long enough to come up with something genuinely interesting?, to envision and mix ideas in different ways, to juxtapose them in unexpected ways, and to "hang in there" mentally when many students' minds want to jump to a more quickly found "good enough" answer so they can move on.

Morgan's solution was a wonderful mix of low tech and high tech—the high tech making a low-tech standard come alive in new ways. It began with students using dry erase boards to brainstorm, outline, and organize their thoughts—a wonderful low-threat method that allows for risk taking and a fluidity of ideas. However, when students felt they had reached a good point in the "discussion" that was coming alive on the dry erase board, they captured it (the whole board or just a great set of comments on one section of it) with a cell phone and posted it to a group shared folder. Being able to capture these otherwise fleeting moments of insight make this temporary medium much more useful—and it is its temporary nature that gets the students to explore ideas more.

For homework, students selected an image, either from their group or someone else's, and added their own text commentary as a textbox overlaid on the image. They then posted this to a class blog, with the additional responsibility to comment on other students' images. These then became the starting point for the next day's class discussion. The ultimate goal of this multiformat, playful exploring of ideas was to create a topic for an essay that they then wrote.

The use of technology provided novelty to help students stay involved in a deep intellectual discussion longer—don't let the "fluffiness" of the methods fool you; with the right guidance from the teacher, they work together to help students wrestle with intellectual ideas for longer, beyond the point where they might otherwise have put them down. The technology also creates moments where different voices will shine and different voices will be challenged. The technology wasn't the focus of this assignment; the teacher kept the focus squarely on the class having a high-quality intellectual discussion, but one that was a rich, inviting, unfolding journey. The technology created more interesting starting points for the essays and more interesting tendrils to explore.

Another way that technology can enhance dialectic is by bringing other voices into the classroom. Video-messaging and video-conferencing technologies can help bring real people into the classroom—as in our example from Haiti. Empathy is an underused tool in teaching, and weaving real people into a class can be a powerful experience.

Technology, then, can play a role in enhancing the dialectic. But it is crucial to remember that the dialectic remains the focus; the quality of thinking that unfolds in the discussion is paramount, the technology is just a tool.

The same is true of the third stage of the trivium, rhetoric—students communicating their thinking. Again, technology is perhaps at its best when it is seen as a tool to enhance the communication. Quality communication is the high bar we are setting our students and should be the stated focus of assignments—the technology is one thing we might use to help us get there. There is an important difference between "make a PowerPoint of a biome" and "use PowerPoint to create a poster that shows your understanding of a biome." Or even, as a student's repertoire of technology skills increases, "create a visual representation that shows your understanding of a biome." They may use any one of a number of technologies, or they may not use a technological solution at all, and that might, for this assignment, be okay.

We just chose the "use PowerPoint to create a poster" example deliberately—to illustrate how technology can allow students to produce work in an iterative fashion rather than one-and-done. Think of how a student would go about writing a history paper. He or she might create an outline, show it to the teacher and get some feedback, edit the outline, show it to the teacher again and get more feedback, tweak it some more, flesh out the outline into paragraphs and create a first draft. Then the student might get feedback on this draft, either from peers or the teacher, and edit it again based on this feedback. The student may do this multiple times until he or she gets the final paper. It is a process full of feedback, reflection, and chances to edit—what we could call an iterative process. We teach students an iterative process for writing so that they can produce a better product.

Now, for example, think of how we typically have students create that other educational staple, a poster. "Okay, students, create a trifold of . . ." And even if they word-process text boxes, editing the whole poster, the entirety of the layout and how everything works together to create a great whole, is hard. The process is far from iterative and often frustrating. Now, many academic disciplines, particularly science, actually use posters as a standard means of communication—if you have ever been to a professional science conference, poster sessions are common.

Making a quality poster, using technology, is a skill that it is worth teaching students. Importantly, it allows us to mirror the iterative history-paper-writing process—chances to get feedback of small versions printed on a single sheet of paper, chances to edit by simply sliding boxes around swapping things in, chances to reflect, edit, reflect some more, edit some more, and make a higher-quality product. Large-format printers are not particularly expensive to buy or run, or the final poster show could be the teacher showing them on a projector or posting them as an electronic gallery—"How do the students show them?" is not an insurmountable barrier.

Technology not only gives students more chances to edit, it does so in a wide and expanding variety of media. Video, podcasting, infographics, stop-frame animation, electronic portfolios, web pages—there is a rich variety. But are they worthwhile? Or are they just Papert's "sugar coating" gimmickry? MBE research suggests that novelty can enhance deep, reflective cognitive engagement; it also suggests that arts integration may aid motivation, as well as memory storage, and many of these communication technologies involve an art component. This might be the power of some of these communication technologies, but they need to be selected and talked about with care by the teacher so that quality communication of thinking is the clear goal.

The richness of media in which students can communicate also ties in with MBE research on the importance of teaching and assessing in multiple modalities. By varying modalities, not only are we more likely to engage more students and have them use content in ways that will help it "stick" for them, we are more likely to find a pattern where different students are challenged and shine as the class moves on. The modalities should be selected with the content in mind—certainly not based on perceived "learning styles" of students in the class. In the words of Professor Howard Gardner of Harvard University, author of the theory of multiple intelligences:

> It may seem that I am simply calling for the "smorgasbord" approach to education: throw enough of the proverbial matter at students and some of it will hit the mind/brain and stick. Nor do I think that this approach is without merit. However, the theory of multiple intelligences provides an opportunity, so to speak, to transcend mere variation and selection. It is possible to examine a topic in detail, to determine which intelligences, which analogies, and which examples, are most likely both to capture important aspects of the topic and to reach a significant number of students. We must acknowledge here the

cottage industry aspect of pedagogy, a craft that cannot now and may never be susceptible to an algorithmic approach. It may also constitute the enjoyable part of teaching: the opportunity continually to revisit one's topic and to consider fresh ways in which to convey its crucial components.[5]

And while he is talking about presenting material here, the same holds true for the tasks where we ask students to work with and present knowledge—all three parts of the trivium.

As students build a repertoire of technology skills with which to communicate, there is increasing room for giving students choice over how they communicate their thinking. Are these various forms worthy? Think of a great example of an informative and moving one- to two-minute video you have seen. Think of a great infographic you have seen. Think of all the brain skills necessary to create it. Think of how facile you need to be with the underlying knowledge in order to select, organize, and juxtapose it in that way.

Yes, essays have massive worth, and writing them is a mighty important skill to teach. But being able to communicate in other ways is important too—increasingly so if we look at the kinds of jobs our students might one day end up in. Let's teach toward these, and try to do so in ways that are iterative—that allow the feedback, reflection, editing, and structure that we know to be so important in essay writing. This is the heart of the potential for the use of technology in dialectic.

While the trivium gives us a structure for where we might use technology, not everything in this chapter fits into it; for example, multitasking and task switching. Despite what most students may think, brain-imaging studies have shown that there is no such thing as multitasking. Instead, the brain rapidly switches back and forth between the networks necessary for each task. But there is a transaction cost for doing so—both in terms of speed and effort.

Studying is harder and takes longer. Students do not understand the transaction costs associated with switching between tasks. In addition, when they study and then receive an indication of an incoming text, their brains experience a dopamine boost that makes it hard to resist the message. However, when they are done looking at or responding to the message they don't return to their work at the same place they just stopped. In essence, they are extending their study time. However, having a more efficient attitude toward multitasking is something we have had a very, very hard time convincing our students to adopt. Classroom-management software, such as DyKnow, offers tools to mitigate this in class, but what about when class is over? This is a very fertile area for a small group of teachers or a school to take on as a research project.

One of the benefits of technology is that it has led to a democratization of learning. Free resources like Khan Academy and Learnzillion give access to better teaching than many kids will ever get. Stanford Online and MIT's Opencouresware are just two examples of a remarkable movement by some colleges to make, quite frankly, staggeringly high-quality thinking, often packed very well, available to all for free.

One interesting use of technology is to have students make e-portfolios. This may help them keep a body of work with them over a number of school years, and allows them to capture a rich array of assignments in various media. Such e-portfolios can be used as part of reflection and metacognition exercises, learning strategies that are supported by MBE research.

We also want to reemphasize the potential for technology to give students rapid, structured feedback and the ability to fix or build from their mistakes. Great tools exist across subjects and grades—from grammar to math. The best often include clever hooks, gaming or rewards systems to keep students engaged. But the best also include something that isn't so visually obvious—they are adaptive. This means that the tasks students are set depends on their prior performance. Too easy? Let's put some harder things in, maybe skip this next section even. Got this question wrong? Let's put another similar one in later and see if it was a mental slip or a gap in understanding. Let's also provide some scaffolded feedback for them to work through. If there is a gap in understanding, let's head off and do *this* unit and *this* unit.

The goal is to match the challenge of the task to the skill level of the student—and to continually adapt to keep the student in the "zone of proximal discomfort" (see chapter 7, "I Love Your Amygdala!") or Mihaly Csikszentmihalyi's zone of flow, as shown in the diagram below.[6] Adaptive software counters the twin demons of boredom and anxiety. Writing adaptive software is harder than writing a one-size-fits-all program, but the learning benefits in many scenarios are so vast that it should be the new goal.

As we ponder future developments in the capacity and cost of, among other things, computer memory and processing power, we may predict that systems with robust performance will become even cheaper and more ubiquitous. Given the potential for adaptive student technologies to personalize learning, differentiate instruction, and increase student engagement and achievement, we might imagine their use becoming significantly more widespread in the near future. However, exactly how this

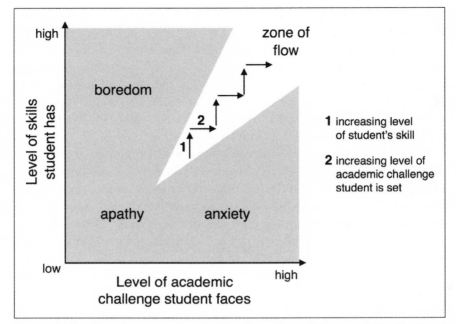

Figure 11.2. Keeping a student in the flow channel by matching the level of challenge of the assigned task to the developing skill level of the student. Academic growth occurs through a coordinated raising of challenge level and skill level in order to avoid anxiety, boredom, or apathy. Adapted from *Flow: The Psychology of Optimal Experience* by Mihaly Csikszentmihalyi (HarperPerennial, 2008), and *Finding Flow: The Psychology of Engagement with Everyday Life* by Mihaly Csikszentmihalyi (Basic Books, 1998).

potential is best translated into actual increased engagement and achievement is, at this time, an unfolding story. But it is one that promises to be interesting to watch.

The role of technology in STEM or STEAM also needs to be addressed, although this is not the place to fully discuss the merits or otherwise of these acronyms. We feel they are limiting—that the ultimate goal is the use of components pulled from science, technology, the arts, math—but also from the humanities, social sciences, and whatever else is needed—to solve problems that people have. This is design thinking—and STEM or STEAM is *one* of the tools our students use to solve both real and contrived or constructed problems that have their roots in people and the real world.

The main point of this is a familiar one in this chapter—that technology, be it modeling, programming, using Excel, or any one of a large and expanding repertoire of skills, is a tool involved in a greater task. Technology is best when it is being used in a greater context. Our students create, amongst other things, automatic plant-watering systems, artificial limbs, and devices and systems to improve the quality of life in Haiti's Central Plateau. The hooks of whimsy or empathy and the contexts of these projects create a drive for learning different technologies—students NEED these technologies to make their "baby," their idea that they have personal attachment to, work.

The essential question when exploring the value added of technology to teaching and learning for us at St. Andrew's has been, how can this hardware or software deepen learning? But with so little research still available, what should teachers do? In the case of one teacher, he did not wait for the empirical research study. In a one-to-one laptop school, he wanted to know whether students taking notes in the traditional way, by hand, were able to retain more of what they wrote than students taking notes using a laptop. The results were compelling, as the students who took the notes by hand retained 40 percent more than those who took the notes by laptop.

Since then, research has validated this teacher's independent research.[7] We include this story to inspire you. The use of technology to enhance learning is an excellent candidate for the type of teacher research we discuss in chapter 12. For example, there has been a lot of discussion of the differences of students reading online or from a printed source.[8] What is best?

We believe strongly that providing teachers with frameworks leads to quicker adjustments in their professional practice and a way to measure their instructional changes. Two additional frameworks we have found useful are Technological Pedagogical Content Knowledge (TPACK) and the Substitution Augmentation Modification Redefinition (SAMR) model—but both need tweaks.

TPACK (TECHNOLOGICAL PEDAGOGICAL CONTENT KNOWLEDGE)

TPACK is based on the Venn diagram below. It is an interesting starting point, but we argue that technology is not a separate sphere in itself, but rather a part of the pedagogical knowledge and content knowledge spheres. Technology is, undeniably at this point of time, part of pedagogy. For example, Ian uses molecular simulations to help students visualize and think through atomic scale processes like evaporation or chemical equilibria. We need professional development to: (1) expand our knowledge of the technologies that help us teach our subjects (and do so in light of MBE research) and (2) expand our knowledge of effective pedagogy where the

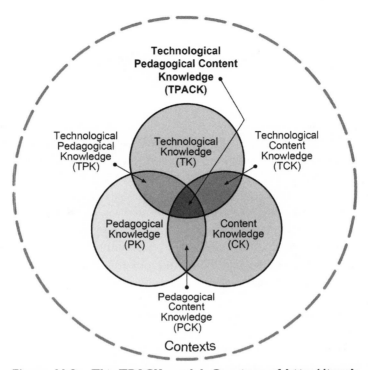

**Figure 11.3. The TPACK model. Courtesy of http://tpack
.org. Reproduced by permission of the publisher, © 2012 by
tpack.org.**

teacher is NOT the absolute master of the technological knowledge, but
effectively manages the class using the knowledge that exists in the room.

Technology is also part of content knowledge because our subjects
have shifted, and will continue to shift, in the perhaps decades since we
studied them in college. How current are we? Part of staying up to date
in content knowledge is learning the technologies that are authentically
germane to that discipline—including in the professional practice of that
discipline and the higher level learning of that discipline. For example, as
a teacher of various courses in the physical sciences, Ian includes modeling
and programming—whatever his comfort level with them, they are NOW
very much a part of his discipline and should be taught. Likewise, Glenn, a
history teacher, includes instruction on how to use electronic citation data-
base software. Which raises the question, how do we structure professional
development to help us so that our kids don't just end up learning what we
did and the way we did it decades ago?

This intersects with an idea we will discuss further in chapter 12, "Teachers Are Researchers," where we introduce the "instructional core" triangle diagram, figure 12.1, which highlights areas for teacher professional development and research. We need, as teachers, to work on professional development on the teacher–content relationship and the teacher–student relationship—technology features in both.

In our examples of molecular simulations, modeling, programming, and using citation software, the technology is not the point of the lesson. They are tools we are using to address a content-specific topic. Technology is used in a greater context. Which gets us back to removing the technology circle from the Venn diagram. This is a significant statement—technology should rarely be the focus of teaching, it is a tool to enhance learning.

THE SAMR (SUBSTITUTION AUGMENTATION MODIFICATION REDEFINITION) MODEL

SAMR defines four levels of technology integration:

Redefinition—Technology allows for the creation of new tasks previously inconceivable
Modification—Technology allows for significant task redesign
Augmentation—Technology acts as a direct tool substitute with functional improvement
Substitution—Technology acts as a direct tool substitute with no functional change

While this can help teachers imagine ways they might include technology, it can be hard to resist thinking of it in hierarchical terms, with "redefinition" as the supreme ruler that we should continually aim for. We just don't think this is always true. The best SAMR level of technology use will vary from task to task—so think of SAMR in the context of Howard Gardner's earlier "smorgasbord" quote. What (if anything at all for that matter), is going to best enhance learning of *this* topic?

In an analogy to ice cream, the SAM levels are vanilla: Regular, French, and Madagascar Bourbon. The R level is Rocky Road (with, maybe, additional caramel sauce, whipped cream, and sprinkles). Vanilla comes in many forms, but all with a certain essential vanilla-iness. And at times this is exactly what is needed, often to enhance another morsel—Rocky Road is too much. And at other times, a full-blown Rocky Road sundae hits the spot. So maybe a better way is to think of this as the SAM–R–NO model. Use your knowledge of your subject and pedagogy and pick what you need, or none at all, to enhance learning. In an unfolding world of shifting and omnipresent technology, this is perhaps a good model for the modern professional craft of teaching.

As an interesting final aside, according to an article in the *New York Times*, Silicon Valley executives—people who live at the boundaries of new technological wonders every day—often choose to send their children to Waldorf and other schools that are technology-less or technology-light[9]; schools where there is a focus on engagement that comes with human contact—where the quality of human contact between students and their teachers and peers is treasured and nurtured to the point where the unnecessary use of technology is seen as a distraction. Are these students missing out?

As one father of a fifth-grader, who is an executive for Google, pointed out, what's the rush to learn these skills? *"It's supereasy. It's like learning to use toothpaste . . . At Google and all these places, we make technology as brain-dead easy to use as possible. There's no reason why kids can't figure it out when they get older."*

Also layered into the technology-learning Dobos torte cake are studies on the amount of screen time children are now experiencing and what detrimental effects this might have. Recent studies suggest that the average eight- to ten-year-old has nearly eight hours of screen time per day—a mixture of TV, computer, and phone—while teenagers have more than eleven hours per day.[10]

The reality of screen time is very much at odds with the guideline from the American Academy of Pediatrics (AAP), which recommends one to two hours per day. By this point in the book you should know that doing anything for an extended period of time will begin to rewire the brain. So imagine what is happening to the brains of students as a result of their prolonged screen time. We know there is still a lot to learn about how the brain learns, works, and changes. We also know that there is still a tremendous amount to learn about the impact technology is having on students' brains, but we do know that it is having an impact, such as affecting children's ability to read emotions.[11]

As schools move more and more tasks to computers, routine tasks like word processing, researching on the Internet, and accessing course and school information on a learning management system mean that the AAP's one- to two-hour recommendation can be easily met without even considering any social or entertainment screen time. The transformative potential of technology versus medical guidelines—this tension alone is worthy of discussion, and is a further call for teachers, school leaders, and policy makers to be very thoughtful and selective in where they choose to include technology to enhance learning.

Technology does, however, have a deserving place in education, and the when, where, and how of it is one of the interesting stories of our time. Who knows where technology will be in twenty or thirty years' time. But we are bold enough to so say that technology alone can never replace the teacher. Relationships remain key.

Without looking back from this page, what are the *three* most salient points you take away from this chapter of *Neuroteach*?

What are *two* things you would like to do "tomorrow" with the information you learned from reading this chapter?

What is *one* question you have after reading this chapter?

NOTES

1. Seymour Papert, *Afterword: After How Comes What: The Cambridge Handbook of the Learning Sciences* (Cambridge: Cambridge University Press, 2006).

2. Seymour Papert, "What's the Big Idea? Toward a Pedagogy of Idea Power," *IBM Systems Journal* 39, no. 3/4 (2000): 720–29.

3. Daisy Christodoulou, *Seven Myths about Education*, first edition (London and New York: Routledge, 2014).

4. Martin Robinson, *Trivium 21c: Preparing Young People for the Future with Lessons from the Past* (Bancyfelin, UK: Independent Thinking Press, 2013).

5. Howard Gardner and Charles M. Reigeluth, eds., *Instructional-Design Theories and Models: A New Paradigm of Instructional Theory*, first edition, vol. 2 (Hillsdale, NJ: Lawrence Erlbaum Associates, 1999).

6. Mihaly Csikszentmihalyi, *Flow: The Psychology of Optimal Experience* (New York: Harper Perennial, 2008).

7. Pam A. Mueller and Daniel M. Oppenheimer, "The Pen Is Mightier Than the Keyboard: Advantages of Longhand over Laptop Note Taking," *Psychological Science*, April 23, 2014, 0956797614524581, doi:10.1177/0956797614524581.

8. Andrew Dillon, "Reading from Paper versus Screens: A Critical Review of the Empirical Literature," *Ergonomics* 3510, no. 10 (2007): 1297–326, doi:10.1080/001401 39208967394.

9. Matt Richtel, "At Waldorf School in Silicon Valley, Technology Can Wait," *New York Times*, October 22, 2011, http://www.nytimes.com/2011/10/23/technology/at -waldorf-school-in-silicon-valley-technology-can-wait.html.

10. Council on Communications and Media, "Children, Adolescents, and the Media," *Pediatrics* (October 28, 2013): 2013–656, doi:10.1542/peds.2013-2656.

11. Yalda T. Uhls, Minas Michikyan, Jordan Morris, Debra Garcia, Gary W. Small, Eleni Zgourou, and Patricia M. Greenfield, "Five Days at Outdoor Education Camp without Screens Improves Preteen Skills with Nonverbal Emotion Cues," *Computers in Human Behavior* 39 (October 2014): 387–92, doi:10.1016/j.chb.2014.05.036.

⑫

TEACHERS ARE RESEARCHERS

> Know thy impact.
>
> —John Hattie

Teachers are researchers. They collect enormous amounts of data each day, and they rapidly evaluate and make decisions based on those data. Some of this is numerical, but much is qualitative. They may be second only to doctors in doing this. What teachers are not good at is doing anything formal with these data. Like a cup sitting under a running tap, more and more information is constantly flowing in, too much to hold. All these data could be used as evidence to inform practice.

Teachers have long been involved in the field of education research. Ironically, however, it is almost always as the subject—a fish inside a bowl—and rarely as, or in collaboration with, the researcher. Again, this is a missed opportunity to inform practice.

Teaching as a profession is also notoriously bad at embedding professional development in the core of how it operates. A review by the Teacher Development Trust found that only 1 percent of continuing professional development delivered to teachers was high quality. Our students spend more than one thousand hours a year learning to get better at their core knowledge and skills; it is rare that a teacher will get above thirty hours a year.

Problems of quality and allotted time exist. A comparison to the medical world is interesting, and so here is a story of a doctor we know who is a radiation oncologist at a major New York City hospital. It is a high-pace, high-stakes department with time pressure to see patients and give them high-quality time and care. Given this time pressure, how do you think they start each day? With a one-hour professional development talk or workshop.

The importance of being current with research, the importance of all people on the team at all points in their professional careers feeling as if they are keeping at the cutting edge of a dynamic field, means that precious time is put aside to do this. This is important time, and it needs to be scheduled in. So a schedule is created that gives value to having up-to-date research, and best practices informed by this research at the core of what you do. This is what it means to be a professional. By this standard, teaching does not do well. At a workshop we gave we heard the phrase, "schedule what you value," which may be an interesting starting point to evaluate your own school.

Imagine a school that put some regular time aside for high-quality professional development, and made it a core part of who they were and what they did. Lucy Crehan,[1] who has studied, worked with, and lived with teachers in countries all over the globe that are widely considered to have great education systems, notes that time put aside for collaborative professional development is a common feature of these schools. It does not have to be every day, but it does have to be frequent and regular.

Let us think about the role of the teacher here. The most important factor is not their being a consumer of information on up-to-date, research-informed methodology, good though this is, but rather their role in being the providers of this information, based on collaborative work with their colleagues. It encourages reflective iterative practice, informed by research and done in collaborative fashion. This is high on the list of "most important things to do" for teaching as a profession that needs to professionalize itself.

To aid this there is a need to create routine, low-threat, easily doable opportunities for teachers to present research-informed methodology. Some teachers occasionally present at conferences, and while this is good, it is not good enough; it is the wrong type of scale for widespread use. Presenting at a conference of five thousand people with thirty concurrent sessions, you may well have only fifty people in the room—an audience size that most schools themselves could exceed. Reflective, iterative, collaborative, research-informed practice, which is then disseminated, is one of the factors that characterizes professional professions—it is time we put it at the heart of teaching.

The research could be a classroom translation or application of methodologies outlined in, preferably peer-reviewed, literature, or it could be novel research conducted by the teacher or cohorts of teachers in a school, or a combination of both. This is a little different than the standard teacher-conference presentation, which tends to be "this is very cool/effective/both, let me show you what I do." Most importantly, this is not how teachers' brains and practice change.[2] It needs to go further than this—this is the research behind *why* this works; or this is the qualitative and/or quantitative data I got from my class that illustrate *how* it is working.

One reason for doing so is that academic research only takes a teacher so far—it gives good avenues, good strategies to explore, but exactly what these look like in the context of any particular classroom is work for teachers, individually or in small groups, to figure out. Doing so and sharing what you find is good work.

Another reason for doing so is that teaching needs to be more research-informed—especially at this time when research exists on how learning might best occur. One of our favorite riffs from Carl Hendrick is the analogy of a doctor who still uses leeches to treat patients, and when questioned on it, holds up his or her hands and says, "works for me." This is not good enough.[3]

As Professor Rob Coe at Durham University in the United Kingdom and the Centre for Evaluation and Monitoring, an important and well-spoken advocate for finding and using evidence to help inform decisions in education, said, "The problem with what's obvious is that it is often wrong." There are many leeches in the standard teaching canon, many preconceptions that do not hold up to evidence. Let us build a profession on something that is more solid than leeches.

Oral presentations, research poster sessions, written publications, online media—the format of the research presentation does not matter as much as the transformative nature of the process. Teachers doing research—or, if you prefer, reflective iterative practice—is the key principle. Three elements elevate this: cohorts of research teachers in school to collaborate together, support each other, and help provide a corner of creative buzz; access to scheduled, regular opportunities to disseminate their findings to a wider audience to aid motivation; and access to a mentor, either in person, online, or both, to play the professor-to-a-grad-student role, aiding with both research process and content knowledge. All these factors help embed research as a routine part of the profession of being a teacher, getting us closer to the medical model of what it means to be a professional in a dynamic field where being at the cutting edge of what is known makes a critical difference.

Who organizes this? Who helps teachers find relevant research literature to bring into their work? Who helps teachers create research processes that are both workable and will provide meaningful results? Who helps guide teachers to find the meaning of the results they gather? Who helps be the bridge between the worlds of academic research and classroom practice? It creates an interesting role in a school: "Head of Research," a title first introduced to us by Carl Hendrick, Head of Research at Wellington College, an independent eighth- to twelfth-grade school in the United Kingdom, or "Research Lead," a more common title that is starting to emerge, predominantly in secondary schools.

What a great role to which a teacher can aspire! Think of the skills and knowledge a teacher needs to build to fulfill this role, and the things he or she could do. Think of the possibilities for this role in a school district where you have the opportunity to create networks of educators, spanning disciplines, grade levels, and schools, who want to be at the cutting edge of a dynamic field.

Who supports this? There is a potential here to build two-way relationships between universities and other foundations or trusts engaged in education research. These institutions gain from an increased opportunity to discover real-world research questions worthy of investigation, as well as having access to schools in which to conduct research. Schools benefit from expertise, gateways into the research literature, and the possibility of having expert research conducted on questions that matter to their mission.

The education world gains from having new ideas and the latest research brought into the realm of classroom teaching, rather than the current pattern that mostly just sees a recirculation of existing ideas. One great example is Research Schools International, a program run by faculty from Harvard's Graduate School of Education, with whom we at St. Andrew's partner. Research partnerships also have the potential for helping schools develop their own protocols for conducting a level of research-informed iterative practice that lies somewhere between peer-reviewed-journal-level research and "reflective practice" that is done in some schools.

Bringing new ideas into education is vital, especially as there are disciplines with ideas that need to be brought in, ideas that will lead to better teaching and better learning for all students. For example:

> The importance of neuroscience in education is becoming widely recognized by both neuroscientists and educators. However, to date, there has been little effective collaboration between the two groups, resulting in the spread of ideas in education poorly based on neuroscience.[4]

There is an opportunity here to embed a culture in teaching that has the power to transform professional practice.

THE FOCUS OF TEACHER RESEARCH

What could this research focus on? We suggest two strands: (1) Teaching strategies informed by mind, brain, and education (MBE) science; and (2) curriculum understanding.

This idea fits with another beacon of education theory, the "instructional core" proposed by Elmore et al.,[5] which discusses how the quality of instruction depends on the relationships among teacher, content, and student, as shown in the diagram below. But focus on the lines, not the circles. The really important, and perhaps counterintuitive, point to note is that it is the relationships between pairs that is the most important element—the lines—rather than the quality of each individual component part—the circles.

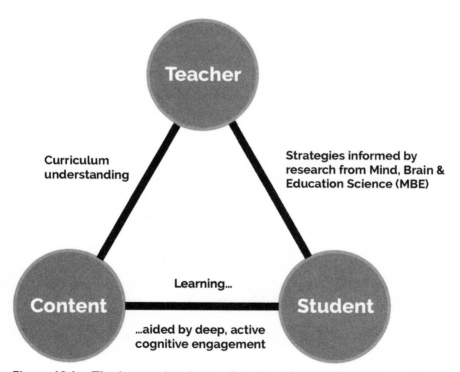

Figure 12.1. The instructional core. Based on Richard Elmore's "Instructional Core" and adapted from Elizabeth A. City et al., *Instructional Rounds in Education: A Network Approach to Improving Teaching and Learning* (Harvard Education Press, 2009).

The critical relationship that all teachers are trying to impact is the one between the student and the content—*this is learning*. In order to do this, our teacher research model works to strengthen the two relationships that the teacher can directly control:

1. Teaching strategies informed by MBE science: strengthens the teacher–student relationship
2. Curriculum understanding: strengthens the teacher–content relationship

To improve learning, professional development must focus on *both*. In addition, the teacher–student relationship stresses the importance of teachers really knowing each of their students—their current strengths, their current weaknesses, their developing voice, and the stories that each of them bring to the class. Growing these skills should be a focus of professional development too.

TEACHING STRATEGIES INFORMED BY MBE SCIENCE RESEARCH STRAND

Published research on MBE science has suggested many strategies to improve teaching and learning, but often the next-level-down details of what these exactly look like in a particular context, the essential "how do these work best in our classrooms?" question, need to be fleshed out. This is fertile ground for teacher research. It is of particular importance since there must inevitably be some degree of subject, age, and school variation in how strategies are implemented. For example, what active retrieval methods work to aid memorization in history? And how are they different from active retrieval methods that work well for biology? What methods work well for eighth-graders as opposed to twelfth-graders? What works in *this* school? In *this* classroom?

For example, "Know Thy Impact" is the title of an article by prominent educational researcher John Hattie. In it, he discusses feedback, a factor that he found had one of the highest effects on student learning in his study of more than nine hundred meta-analyses of educational research. However, this is his first sentence: "Teachers give a lot of feedback, and not all of it is good." He goes on to say, "The effects of feedback, although positive overall, are remarkably variable. There is as much ineffective as effective feedback." What does effective feedback look like in the context of your classroom? What actually is your evidence that you are indeed giving effective feedback? This is a great teacher research avenue to explore.

Such teacher research may change individual practice, but it may also create momentum to shift practice and grow cohorts within schools or districts, as well as having the potential to impact the direction of future academic research in the field of MBE science. In our first research strand for teacher researchers, teaching strategies informed by MBE science, we suggest the following research questions to explore:

1. Ready-made research, from our current list of favorite authors. Investigate how you can use their research to inform your practice.

Angela Duckworth	Paul Howard Jones
Carol Dweck	Eric Kandel
Kurt Fischer	Mark McDaniel
Howard Gardner	Michael Posner
Jay Giedd	Todd Rose
John Hattie	Tracey Tokuhama-Espinosa
Christina Hinton	Daniel Willingham
Mary Ellen Immordino-Yang	Judy Willis

2. Teacher research. Table 12.1 contains research topics informed by MBE science that have solid evidence to support them. The goal is for teachers to take one of these research areas and investigate the use of practices that work in their particular context. Then, with a spirit of altruism, disseminate this knowledge to help others find ways to use it in their own, different context.

Table 12.1. Twelve Possibilities for Teacher Research

Loci

i.	Using feedback.
ii.	Using formative assessments.
iii.	Using metacognition.
iv.	Explicitly teaching memorization strategies, such as active-retrieval methods.
v.	Providing time and methods for student self-reflection.
vi.	Providing students with choice.
vii.	Building a growth mindset through the use of specific learning strategies and reflection on how they are working.
viii.	Using peer tutoring.
ix.	Providing students with explicit instruction on how to plan and organize in order to help them enhance their executive functioning skillset.
x.	Providing students with experiences and environments that aid identity validation, and eliminating factors that may cause or enhance identity threat.
xi.	Integrating the visual and performing arts into non-arts subjects to create knowledge transfer.
xii.	Using play constructively to increase student engagement.

This table is by no means complete—but it is a good place to start for two reasons. First, the research supporting these factors is good. Second, the possibilities have promise—using these as starting points, small groups of teachers working together should be able to find strategies that improve learning in their classes. For other research possibilities we have looked at, check this footnote.[6]

There is also an interesting angle to this research strand that MBE science is just beginning to explore, and which would make for wonderfully ambitious teacher research. That is the question of conducting research on teachers. For example, what kinds of professional development lead to increased use of MBE research-informed strategies in ways that lead to higher student achievement? This has the potential to be truly great work.

3. Original/university-supported research. These are topics we are investigating at St. Andrew's, where support from a research institution has been vital to collect and analyze data in a meaningful way.

> How do peer relationships impact stress and student achievement?
> How does happiness shape student motivation and academic achievement?
> How does MBE science currently inform teacher, student, and parent practices?
> How can increased training in MBE science improve teaching, professional satisfaction, and student achievement?

As well as enabling us to conduct research with a higher level of academic rigor, these collaborations greatly enhance the level of knowledge and expertise available to us, which benefits all the teacher research that is going on at the school. It also inspires us to stay curious, raise our game, and make teacher research part of who we are as a school. In the other direction, the connection with schools leads academic researchers to craft research studies that have direct implications to schools.

CURRICULUM UNDERSTANDING RESEARCH STRAND

There is a fear that generalist approaches to MBE-informed strategies, strategies that do not take the particular demands of different subjects or grade levels into account, will fall flat. The effective teaching of any specific subject, say history, demands such a detailed understanding of how knowledge in that subject is constructed that teaching strategies that do not take this into account will be flawed. This is an argument eloquently put by Michael Fordham, a senior teaching associate in the Faculty of Education at the University of Cambridge. We include his thoughtful list of curriculum questions that each subject teacher "need[s] to ask in order to understand what it is that we are teaching" below, plus two of our own [in brackets].

> So what might it mean to think about the subject? What are the kinds of curriculum questions that we need to ask in order to understand what it is that we are teaching? I would suggest some of the following are important to nearly all subjects:
>
> - [What are the essential representations of my subject?]
> - How can knowledge in a subject be structured?
> - What happens to knowledge in the process of teaching?
> - What are the structures of the concepts that we teach?
> - How is knowledge of one thing dependent on knowledge of another?
> - What are valid questions to ask in a subject?
> - How do certain ideas gain and lose meaning in the process of generalization?
> - What are the limits of what a subject can explain?
> - What would it mean to use a subject inappropriately?
> - What can be assessed in the subject, and in what ways?
> - What does it mean to get better at a subject?
> - [With what concepts do students typically struggle, and in what ways can I help them to reach necessary understanding?]
> - In what ways can one subject interact with another, and in what ways can it not?

Teacher research into answering these questions is also a great research pathway to take. They are not questions that an undergraduate degree in a subject typically forces you to think about, yet they are critical for effectively teaching a subject. So how do we purposefully create space for teachers to explore and research these questions? This is where our second suggested research strand for teacher researchers comes in: "curriculum understanding."

However, deepening teachers' understanding of their respective subjects is not mutually exclusive from using MBE science to inform practice; an expert teacher can, and should, do both. Imagine teachers whose curiosity and passion for their subject drives them to reflect deeply on the points in

Fordham's list above; but they are also someone whose passion for their craft of teaching drives them to learn about MBE-informed strategies.

We argue that such teachers will also have the talent and drive to reconcile these two strands to figure out what are the best teaching strategies that will work in *this* subject (and with students of *this* age). That is the real power, and the real need, of teacher research. The melding of these two strands at the pointy end of what this actually looks like in a classroom is something that cannot be effectively done by cognitive psychologists or even educational researchers. It needs teacher researchers.

John Hattie, professor of education and director of the Melbourne Education Research Institute at the University of Melbourne, in his 2003 paper "Teachers Make a Difference: What Is the Research Evidence?" presented to the Australian Council for Educational Research Annual Conference on Building Teacher Quality, wrote about the difference between expert teachers and experienced teachers. As a profession, we unfortunately tend to conflate the two. We tend to believe, usually in spite of a number of examples from our own personal experiences of schooling, that experience bestows expert status. That, of course, is not the case.

Hattie outlines key differences, which we expand upon more in chapter 13, where we introduce a pathway for professional growth informed by MBE science. The interesting thing to note is that when you look at the characteristics of expert teachers that Hattie describes, it is a list to which our two research strands—MBE-informed strategies and curriculum understanding—unerringly point. Teacher research, then, may be seen as a pathway to becoming an expert teacher.

CURRICULUM UNDERSTANDING + TEACHING STRATEGIES INFORMED BY MBE SCIENCE = PEDAGOGICAL CONTENT KNOWLEDGE

Thinking back to the comparison to the role of research in the field of medicine with which we started this chapter, let's consider the path of professional growth that teacher researchers are walking down. First, strengthening their pedagogical knowledge through increased knowledge of general MBE science research and how it informs subject-specific teaching, assessing, and learning strategies. Second, strengthening their content knowledge through research in curriculum understanding. Third, these two research strands work together to strengthen *pedagogical content knowledge*.

Pedagogical content knowledge is a type of knowledge that is unique to teachers, and is based on the manner in which teachers relate their pedagogical knowledge (what they know about teaching) to their subject matter knowledge (what they know about what they teach). It is the integration or the synthesis of teachers' pedagogical knowledge and their subject matter knowledge that comprises pedagogical content knowledge.[7]

Kathryn F. Cochran, James A. DeRuiter, and Richard A. King[8] revised Shulman's original model to include *four* major components: subject matter knowledge; pedagogical knowledge; teachers' knowledge of students' abilities and learning strategies; and teachers' understanding of the social, cultural, and physical environments in which students are asked to learn, as shown in the diagram below.

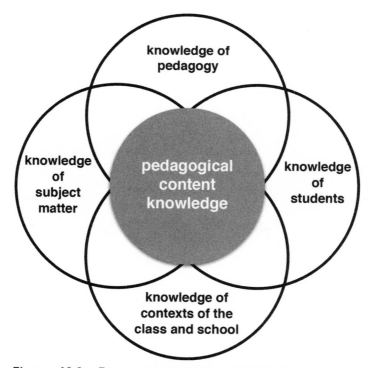

Figure 12.2. Four components of pedagogical content knowledge. Adapted from Kathryn F. Cochran, J. A. De-Ruiter, and R. A. King, "Pedagogical Content Knowing: An Integrative Model for Teacher Preparation." *Journal of Teacher Education* **44 (1993): 263–72.**

Interestingly, this may be seen to fit our two research strands even better, as Cochran et al.'s "knowledge of pedagogy," "knowledge of students," and "knowledge of environmental contexts" may be thought of as three pillars of MBE research. The path of professional growth that our teacher researcher is walking down, therefore, seems to be a good one. This is why the teacher-researcher model should be created, and in those rare enclaves where it has already been created, should be fortified, since it is a pathway to important professional growth, bringing cutting-edge ideas from the research world into areas that have already been identified as being critical to the development of expert teachers. If we can make this possible, doesn't that sound like an interesting thing to do, something that might have a chance of helping attract and retain the best and brightest into a career in teaching?

WHAT MAKES GREAT TEACHING?

"What makes great teaching?" This is the title of the "Review of the Underpinning Research" prepared by Coe et al. for the Sutton Trust. They identify six research-backed components of great teaching, in the context of discussing how to assess teacher quality.

> This should be seen as offering a "starter kit" for thinking about effective pedagogy. Good quality teaching will likely involve a combination of these attributes manifested at different times; the very best teachers are those that demonstrate all of these features.

The top two factors on their list match our proposed teacher-research strands very well:

1. *Content knowledge* (Strong evidence of impact on student outcomes)
2. *Quality of instruction* (Strong evidence of impact on student outcomes)[9]

Now that we have discussed why teacher researchers are beneficial, and have outlined two strands they should research, what might this research actually look like?

WHAT MIGHT TEACHER RESEARCH LOOK LIKE?

Creating a fair survey that actually measures what you think it might be measuring is hard—it is sophisticated work that requires a particular expertise that needs to be taught, and is usually only done in certain graduate-level courses. Likewise, analyzing data to the point where you can extract publishable meaning from them is a high level skill that takes time and mentorship to master.

It is easy to get Excel to show a graph with a lovely line shooting through a cloud of points, but is there any defendable meaning to this line, and if so, what? We can imagine eventually getting to a point where there are mechanisms for talented and dedicated teacher researchers to learn these skill sets, but this pathway would have to be built. But imagine the possibilities of using research to inform decisions if schools had access to a "research lead" who could lead this work.

So what can teacher researchers do? There are less-sophisticated research protocols that can still yield useful results. Indeed, we at the Center for Transformative Teaching and Learning (CTTL) are developing some— an introductory level of research that is highly accessible, and, for those whose appetite is whetted, a more robust protocol that requires more of the teacher, but that is still very conscious of how it must fit in with all the other demands on a teacher's time. We introduce the former later in this chapter, but the latter is beyond the scope of this current book.

Why do we think teachers can be researchers? Consider this: teachers take in and react to vast amounts of data every day—this is just the nature of the minute-to-minute job of being a teacher. Teachers typically do not record these data. So the first research practice to begin is to journal—have the equivalent of the sacred "lab notebook," where all thoughts, ideas, experiences, and insights are recorded, and where going back to erase or change previous notes is strictly forbidden. (You can update with a new note and cross-reference, but leave the original thought alone!)

Creating the habit of writing down some of the trove of daily data is the crucial beginning point for the teacher researcher. While it is the simplest thing to say, it is also perhaps one of the hardest to implement because it runs counter to the busy, busy, busy nature of the typical teacher's day. Doing so requires a mindset shift that this work is important because it will ultimately help *these* students, *these real faces, in front of me.*

Teachers should not shy away from surveys entirely. They are a quick way to get lots of information on a broad range of points, but they must be used in the context of all the other information a teacher researcher obtains. Teacher researchers must be careful to avoid survey fatigue, which may occur in four ways: the number or frequency of surveys that individual students are asked to do; the length of a survey; the balance of written answers to select-an-answer or ranking questions; and the variety of select-an-answer or ranking questions included. In addition, allowing anonymity is always an important judgment call.

Data collected from student interviews, used in combination with data collected from surveys, are much better than either one on its own. While these may be time-consuming to conduct, the simple act of asking students about things which are integral parts of their daily lives is an incredibly empowering experience—try it!

Comparison trials can give interesting information, but the inevitably small numbers of students in such trials, plus variations between sections and years, mean that results must be judged with care, and always in the context of information gathered in other ways. They must be used alongside a variety of data sources that together build a picture—a picture that the teacher researcher works to decode.

When they disseminate the story that emerges from this research, teacher researchers must be careful to use an appropriate level of uncertainty in their language so as to not overstate their case. They must put out their evidence and be circumspect with their judgments. The whole purpose is not to provide perfect "silver bullet" solutions that other teachers or schools can carry into their classrooms, unpack, and use right away. The purpose of these studies is to guide further work—put your ideas out there, seek feedback, tweak, iterate, and keep refining.

Teacher researchers must also not shy away from research literature—and one way to do this is to use this book and the references within it as a starting point.

Observation by teacher peers can provide valuable feedback, an idea that is supported by research. Coe et al. followed their question "What makes great teaching?" by asking "How could this promote better learning?" and, from their research, suggest a teacher "'knowledge-building cycle'—a feedback loop for teachers—that is associated with improved student outcomes." They suggest an analogy between teacher-learning and student-learning, that if we structure teacher-learning in a similar fashion to how we know students learn best, it can "have a sizable impact on student outcomes."

They suggest a type of formative assessment, where teachers receive low-stakes feedback from classroom observation, and that this be a continuous process that is pitched to the teachers as a professional learning opportunity, rather than an assessment, to help improve outcomes for their students. Furthermore, this model works well when the feedback is from fellow teachers. This marries well with the teacher-research model, which may serve to provide an engaging focus for cohorts of teachers involved in this type of peer-feedback professional development, a type of professional development that research suggests has a significant impact on improving student learning. For example, Coe et al. go on to say that "the [research] literature provides a challenge to the much quoted claim that teachers typically improve over their first 3-5 years and then plateau." The most supportive professional environments saw teachers continue to improve beyond this period, whereas the least supportive led to teachers actually becoming less effective after three years.

One interesting model to adapt for teacher-researcher peer observation is "Jugyou kenkyuu," which has long been used for teacher professional development in Japan. Renamed "lesson study," it is a practice that has been spreading globally in recent years, and there is research to suggest it improves instructional practice and student learning.[10] It involves groups of teachers, over the course of several months to a year, creating lessons, observing each other teach them, and then examining and refining the lessons. The cycle might look like this:

- Defining and researching a problem
- Planning the lesson
- Teaching and observing the lesson
- Evaluating the lesson and reflecting on its effect
- Revising the lesson
- Teaching and observing the revised lesson
- Evaluating and reflecting a second time
- Sharing the results[11]

An interesting but less common reinterpretation of "Jugyou kenkyuu" is "learning study"—and we feel this is the basis of what teacher research should be. Whereas the primary goal of lesson study is increasing the skill of individual teachers, learning study also includes a research component. It involves a group of teachers wrestling with the question, "How can X be taught to students so that they have the best opportunity to learn it?"[12] As they do so, "what is being learned?" is as important a question as "what is the best way to learn it or teach it?" (which agrees well with the two research strands we propose in this chapter: teaching strategies informed by research from MBE science and curriculum understanding). It may be the closest thing in education to clinical research, a fundamental part of translational practice, turning research into everyday procedures, in the field of medicine.

It should rather be described as clinical research (in analogy to medical clinical research). The use of teachers' experiences and tacit knowing in the knowledge-producing process, the iterative process of specification of theory, and the uniqueness of the learning problems among different groups of pupils are central aspects of a particularistic clinical research process. In comparison with lesson study, the learning study is more focused on constructing knowledge concerning objects of learning as well as teaching-learning relations. Teachers are included in the research as interpretative professionals making professional sense of particular educational events.[13]

This is what we are aiming for: teachers solving problems of pedagogy in context, in a collaborative fashion, with a reflective and iterative process—problems that are learning-focused: "What's the best way to teach *this* so that *these* students have the best chance of learning it?" This is what learning study gives us. For our research model, we need to add to this the input of ideas from MBE research, so that the types of problems teachers choose to study and the approaches they hone in on to solve them are informed by research. In addition, the dissemination of a readily digestible "knowledge product" at some point is highly desirable. We also suggest that working in small groups makes practical sense—anecdotally, three people to a group seems to work well.

The "learning study" model is similar to the "instructional rounds" idea proposed by Elmore et al., which draws on the comparisons to the medical field, where "rounds" are a key part of professional development.[14] Key to both these ideas is that peer observation and feedback becomes a culture. We agree, this needs to be a culture, but suggest that it is given some direction—using our two teacher-research strands of, first, teaching strategies informed by MBE science and, second, curriculum understanding. The influx of research, as we mentioned earlier, has the additional vital role of bringing new ideas into the realm of teacher professional development, a practice hitherto dominated by recirculation of existing knowledge.

Our profession needs to be characterized by teachers working together to figure out how students might best learn *this particular idea or skill* in the intricate set of contexts of their particular school setting. Their work needs to be informed by current research in how students learn best. Their work needs to be iterative—try something, get feedback, reflect, tweak, and try again. Their work will be all the better if it is guided and supported by "research leads" who have training and experience in such translational practice, bridging the gap between research and practice so that we can truly claim this work to be "research informed." Their work will be all the better if it is supported by a collaboration with an external research body such as a university, foundation, or trust. Reflective, iterative, research-informed, collaborative practice.

RESEARCH-INFORMED REFLECTIVE ITERATIVE PRACTICE DONE COLLABORATIVELY

So what might teacher research, put simply, look like? Consider the process below, adapted from David Weston, chief executive of the Teacher Development Trust, who has done excellent work on, among other topics, what constitutes exemplary teacher professional development. Ideally this happens under the guidance of a research lead; though if such a person does not exist yet in a school, this enterprise might just help find the right person or people for the job.

- ○ Collect a group of teachers (three are good).
- ○ Talk to students.
- ○ Figure out a problem that needs to be addressed. We suggest beginning with one of the following, taken from the list earlier in the chapter. These are strategies with strong and accessible research support that should benefit student learning, but the details of how they work in your context need to be figured out.

  - Using metacognition to improve learning
  - Using peer tutoring to improve learning
  - Using feedback to improve learning
  - Incorporating student choice
  - Using formative assessments
  - Building a growth mindset
  - Techniques to aid memorization (such as active retrieval, spaced learning, and interleaving)
  - Using novelty or relevancy to create engagement
  - Using the primacy recency effect to structure lessons

- ○ Brief yourself on relevant research.

  - What books or papers should you read? A good place to start is the Education Endowment Foundation's "Teaching and Learning Toolkit," which includes well-summarized literature reviews.
  - Is there a conference you can go to?
  - Identify people who are experts on this issue.

- ▪ If you go to a conference, get the business card of the facilitator. Write to an expert. Endeavor to make a personal contact.

- ○ Knowing what research says, talk together in your research group, observe classes together, and grade together. Get a baseline of how things are, before you try to implement a solution to the problem you identified. Use surveys if desired to get qualitative and quantitative data. Interview students or other faculty members, if desired.
- ○ Determine a solution to the problem you identified.
- ○ Implement this.
- ○ Get feedback on how it worked. Talk, observe classes, and grade with the members of your group. Use surveys or interviews again, if you did so before.
- ○ As a group, assess how well your solution worked. If using statistical methods, do so well within the skill level of your group or research lead.
- ○ Email the experts you previously identified or the facilitator at the conference you attended with your observations and how you think they might pertain to their research. You may not hear back, but people often like to hear about their work in action in real classrooms.
- ○ Refine your solution, based on all the sources of feedback you receive.
- ○ Keep tweaking your solution through this iterative process.
- ○ Disseminate your findings. Even if you do not have a complete answer, you will inform and inspire somebody.

This goes beyond the traditional "reflective practice" model that is done in some schools. But it is unlikely to be sufficiently robust research to submit to peer-reviewed journals (unless it is done in partnership with a research institution). It is research-informed reflective iterative practice done collaboratively, and it sits between these two levels. But is it worthwhile? We argue a resounding yes.

We want reasonable responses to genuine problems in a reasonable timescale. So let's play in the sandbox. As long as we take care to make sure that solutions are informed by research in their inception; as long as we take care to get feedback, reflect, tweak, and try again so that our process is iterative; as long as we take care to make sure we are actually collecting and using evidence, and doing so appropriately, we will be moving in the right direction. The nature of the work, the nature of the collaborations will, we think, if honored by the support of sufficient time and mentorship, help make more joyful teachers—and that helps everyone.

BARRIERS TO TEACHER RESEARCH

In addition to championing its power, we must also consider the barriers to teacher research:

Access to research literature—Access to a swathe of peer-reviewed journals costs so much it is beyond the reach of schools, and rights to use university libraries, particularly online services like journal access, are typically tightly guarded. Google and Google Scholar help, but do not find everything, and tracking down that one reference you want this way takes more time than teacher researchers have to spare.

There is a hint of the type of sea change that is needed: since 2008, researchers who have received NIH funding are required to submit free-access articles to PubMed Central. This, however, covers just a small slice of all articles a teacher researcher might want to access. One solution is to try contacting the author and explain who you are and what you are trying to do. We also recommend books from the authors listed in the table above. In addition, here are some free online resources we have enjoyed:

- The "Teaching and Learning Toolkit" from The Education Endowment Foundation (EEF)
- Mind, Brain, and Education: The Student at the Center Series" by Christina Hinton, Kurt Fischer, and Catherine Glennon, *Mind, Brain, and Education*, March, 2012
- "Neuroscience and the Classroom," online course from Annenberg Learner
- "Applying Science of Learning: Infusing Psychological Science in the Curriculum," Victor Benassi, Catherine Overson, and Christopher Hakala, eds., from the American Psychological Association
- "What Makes Great Teaching?" by Robert Coe, Cesare Aloisi, Steve Higgins, and Lee Elliot Major, from the Sutton Trust
- "Brain and Learning," online course from the NEA Foundation

Mentoring—Someone to play the role of the professor to the grad student. What research leads to follow? What to read? Who to talk to? Also, someone to help prevent the misinterpretation or overinterpretation of peer-reviewed literature from quite a broad range of fields. Education desperately needs this "Head of Research" or "Research Lead" job to exist, and for ways to give these people the training they need. It makes other beneficial tasks and projects eminently more doable. But for it to exist, schools and school districts need to see its value, and graduate schools of education need to understand that something that they may view as being in solely their purview can exist within a school setting.

It takes the right kind of person with the right kind of training, and also with the right kind of connections to the higher education and research world—but when this is done right this person can be a powerful ally in bridging the worlds of research and practice. Without this one thing, everything in this chapter is a significant step more difficult. With it, the world of education starts to look like the last speech bubbles in the glorious run of *Calvin and Hobbes*, "It's like having a big white sheet of paper to draw on! A day full of possibilities! It's a magical world, Hobbes, ol' buddy. Let's go exploring!"

Time—When does the teacher researcher do this work? Maybe if we had the ability to reinvent the profession of being a teacher we would make this a fundamental part of what being a teacher was—conducting research in our classes on practices informed by MBE science research, then presenting these to our peers as part of routine and regular professional development.

Think back to the story of the radiation oncologist with which we started this chapter. One possibility we have just started exploring at St. Andrew's is a new take on annual professional development, where a teacher chooses to replace the traditional one to three classroom visits with a research project. The teacher, department head, and the school's research and development team (in our case, the CTTL) work together to create the project and monitor and guide the teacher's progress through it.

Why does this work? First, to respect a teacher's time, it is important to remove something before adding something. Second, it gives ownership and buy-in to the teacher—using MBE-informed strategies to explore MBE-informed strategies! Just as a great student project leads to deep learning because of the deep level of engagement and empowerment it engenders—teachers just know when this happens—this type of project can lead to great professional development.

Access to collaborators—It is lonely to be the only teacher researcher in a school, and dangerous to work in a bubble. Collaboration is vital, both within subject and grade groups, since effective classroom strategies will have a degree of subject specificity, but also between these groups because commonalities abound, and, the history of research in all disciplines tells us, peeking outside our bubbles will inevitably spark ideas. But how do teacher researchers find collaborators outside of their school? There is huge yet-to-be-mined potential here that should also yield that magic pairing of professional growth and professional satisfaction.

Opportunities to present research—For the teacher-researcher model to work, there need to be opportunities to disseminate research, receive feedback, and build relationships. At St. Andrew's, we have made this happen through three avenues: two volumes of our research publication, *Think Different and Deeply*; an annual in-house research poster session; and an online research blog.

We have found that making teachers act like researchers shifts their mindset. Giving teachers permission to pause and focus intently on one small aspect of their craft is an empowering exercise, as is forcing them to step outside their normal humility and share their work on a more public stage. But how does a larger-scale teacher-researcher dissemination forum exist, because the true power unfolds when the connections stretch far beyond one school? Teaching needs these kinds of opportunities to professionalize our practice.

A teacher has many, many tasks to divide his or her time between. When this type of work is not seen as a chore, it means we are attracting the right sort of people into teaching and that we are adequately supporting them with time, mentorship, and resources. Imagine cohorts of teachers in schools who enjoyed being at the cutting edge of a dynamic field with opportunities to create, explore, and collaborate with like-minded professionals.

If you needed a heart operation and discovered the doctor you went to was still doing things basically the same way as when he or she graduated from med school in 1980, we hope you would turn around and leave. We do not hold teachers to anything close to the same standards of "staying current in your field." Imagine the possibilities if we did. Remember, children's brains have a high degree of neuroplasticity throughout their entire school years. Their brains will rewire based on the environments their teachers create; there is an element of "you just have one shot at this" here, as each child goes through this period of schooling and intense brain development just once. We need teachers who are dedicated to informing their practice by research, staying up to date with current research, and seeing themselves as contributors to an evolving research base.

There has been much talk of "The Finnish Education Miracle" and how to replicate it. It is, of course, no miracle, but reflects the high caliber of people choosing education as a career. The compelling question is, how to replicate it? We suggest a three-pronged approach may make teaching more compelling—being at the cutting edge of a dynamic field; the opportunity and freedom to create; and scaffolding to help support this. Teacher research that connects teachers to the field of MBE science helps this happen.

Professor John Hattie has reportedly said that politicians do not invest in teacher expertise because they cannot see it. Reflective iterative practice, informed by research, that is collaborative and has the goal of disseminating knowledge—making this so common in teaching that you would not even think of commenting on it, would help make teacher expertise visible. In Hattie's own words:

> My role, as teacher, is to evaluate the effect I have on my students. It is to "know thy impact," it is to understand this impact, and it is to act on this knowing and understanding. This requires that teachers gather defensible and dependable evidence from many sources, and hold collaborative discussions with colleagues and students about this evidence, thus making the effect of their teaching visible to themselves and others.[15]

Research tells us that teachers having knowledge of MBE science leads to increased differentiation of teaching and better learning. But there is an important second step. As the wonderful Professor Rob Coe at Durham University made stick in our minds, *that strategy might seem effective at improving learning, "but where is the evidence?"* We can make both these steps happen by creating a model where being a teacher researcher is a normal, everyday thing, and where there are avenues and supports that make this work. We demand this of medical practitioners. We do not demand it of the people in charge of the environments that have a great effect on how our children's brains rewire and develop. This is wrong. Learning would be improved for all students if we did.

Without looking back from this page, what are the *three* most salient points you take away from this chapter of *Neuroteach*?

What are *two* things you would like to do "tomorrow" with the information you learned from reading this chapter?

What is *one* question you have after reading this chapter?

APPENDIX A: WHAT MAKES GREAT TEACHING? SIX COMPONENTS IDENTIFIED BY COE ET AL.

Below we list the six common components suggested by research that teachers should consider when assessing teaching quality. We list these approaches, skills, and knowledge in order of how strong the evidence is in showing that focusing on them can improve student outcomes. This should be seen as offering a "starter kit" for thinking about effective pedagogy. Good quality teaching will likely involve a combination of these attributes manifested at different times; the very best teachers are those that demonstrate all of these features.

1. *(Pedagogical) content knowledge (Strong evidence of impact on student outcomes)*
    The most effective teachers have deep knowledge of the subjects they teach, and when teachers' knowledge falls below a certain level it is a significant impediment to students' learning. As well as a strong

understanding of the material being taught, teachers must also under-stand the ways students think about the content, be able to evaluate the thinking behind students' own methods, and identify students' common misconceptions.

2. *Quality of instruction (Strong evidence of impact on student outcomes)*
   Includes elements such as effective questioning and use of assess-ment by teachers. Specific practices, like reviewing previous learning, providing model responses for students, giving adequate time for practice to embed skills securely, and progressively introducing new learning (scaffolding) are also elements of high-quality instruction.
3. *Classroom climate (Moderate evidence of impact on student outcomes)*
   Covers quality of interactions between teachers and students, and teacher expectations: the need to create a classroom that is constantly demanding more, but still recognizing students' self-worth. It also involves attributing student success to effort rather than ability and valuing resilience to failure (grit).
4. *Classroom management (Moderate evidence of impact on student outcomes)*
   A teacher's abilities to make efficient use of lesson time, to coordi-nate classroom resources and space, and to manage students' behavior with clear rules that are consistently enforced, are all relevant to maxi-mizing the learning that can take place. These environmental factors are necessary for good learning rather than its direct components.
5. *Teacher beliefs (Some evidence of impact on student outcomes)*
   Why teachers adopt particular practices, the purposes they aim to achieve, their theories about what learning is and how it happens, and their conceptual models of the nature and role of teaching in the learning process all seem to be important.
6. *Professional behaviors (Some evidence of impact on student outcomes)*
   Behaviors exhibited by teachers such as reflecting on and develop-ing professional practice, participation in professional development, supporting colleagues, and liaising and communicating with parents.

NOTES

1. Lucy Crehan, *Cleverlands, Inside the World's Best Classrooms* (forthcoming).

2. P. Cordingley, S. Higgins, T. Greany, N. Buckler, D. Coles-Jordan, B. Crisp, L. Saunders, and R. Coe. "Developing Great Teaching: Lessons from the International Reviews into Effective Professional Development," 2015, http://dro.dur.ac.uk/15834.

3. Some of you may point out that leeches are actually still used in modern medicine—for instance, they are one of the best ways to remove congested blood in reconstructive surgery. We feel this adds to our analogy. Parts of the classic canon of teaching, like lecturing, for example, have their place in the world of research-informed teaching, but they need to be used in the right context, and in conjunction with other methods—just like the leeches.

4. Eleanor J. Dommett, Ian M Devonshire, Carolyn R Plateau, Martin S Westwell, and Susan A. Greenfield, "From Scientific Theory to Classroom Practice," *The Neuroscientist: A Review Journal Bringing Neurobiology, Neurology and Psychiatry* 17, no. 4 (2011): 382–88.

5. Elizabeth A. City, Richard F. Elmore, Sarah E. Fiarman, and Lee Teitel, *Instructional Rounds in Education: A Network Approach to Improving Teaching and Learning* (Cambridge, MA: Harvard Education Press, 2009).

6. Using novelty or relevancy to create engagement; the primacy recency effect; the importance of relationships; getting students to know about and believe in brain plasticity; ways to increase intrinsic motivation; teaching creativity; stress and learning; what is excellent homework?

7. "NARST: Publications: Research Matters—To the Science Teacher," www.narst .org (accessed December 2014).

8. Kathryn F. Cochran, James A. DeRuiter, and Richard A. King, "Pedagogical Content Knowing: An Integrative Model for Teacher Preparation," *Journal of Teacher Education* 44 (1993): 263–72.

9. We include the full list of six components in an appendix at the end of this chapter as they present a potential interesting spark for possible teacher-research studies to fill in the gaps of what these things might actually look like in classrooms.

10. C. Lewis, R. Perry, and A. Murata, "How Should Research Contribute to Instructional Improvement? The Case of Lesson Study," *Educational Researcher* 35, no. 3 (2006), 3–14.

11. E. C. Cheng and M. L. Lo, "Learning Study: Its Origins, Operationalisation, and Implications," *OECD Education Working Papers*, no. 94 (Paris: OECD Publishing, 2013), doi: http://dx.doi.org/10.1787/5k3wjp0s959p-en.

12. Cheng and Lo, "Learning Study: Its Origins, Operationalisation, and Implications."

13. I. Carlgren, "The Learning Study as an Approach for Clinical Subject Matter Didactic Research," *International Journal for Lesson and Learning Studies* 1, no. 2 (2012): 126–39.

14. Elizabeth A. City, Richard F. Elmore, Sarah E. Fiarman, and Lee Teitel, *Instructional Rounds in Education: A Network Approach to Improving Teaching and Learning* (Cambridge, MA: Harvard Education Press, 2009).

15. John Hattie, *Visible Learning for Teachers: Maximizing Impact on Learning* (New York: Routledge, 2012).

13

FROM RESEARCH TO PRACTICE

All men dream: but not equally. Those who dream by night in the dusty
recesses of their minds wake in the day to find that it was vanity: but the
dreamers of the day are dangerous men, for they may act their dreams
with open eyes, to make it possible. This I did.

—T. E. Lawrence

How important is good teaching? By means of a large statistical metastudy
of over five hundred thousand studies, John Hattie investigated the relative
importance of factors responsible for variances in students' achievement,
with the aim to allow us to *"concentrate on enhancing these sources of
variance to truly make the difference."*[1] Unsurprisingly, he found that it is
students and "what [they] bring to the table" that is the greatest factor influ-
encing academic achievement, accounting for about 50 percent of the variance of achievement. Home, schools, principals, and peers all have small
effects accounting for less than 10 percent of the variance of achievement.
The second-place factor, responsible for about 30 percent of variance in
achievement, is teachers. In Hattie's words, *"It is what teachers know, do,
and care about which is very powerful in this learning equation."*

Intrigued by these initial findings, Hattie went on to explore what it was
that made excellent teaching. He made an important distinction that is
bound to cause some waves: there is a difference between expert teachers and experienced teachers. Experience does not automatically beget

expertise, and expertise does not solely come about through vast amounts of experience. Hattie was interested in understanding "the expertise that underpinned the expert teachers," and identified the following five dimensions of expert teachers:

i.   Can identify essential representations of their subject
ii. Can guide learning through classroom interactions
iii. Can monitor learning and provide feedback
iv. Can attend to affective attributes
v.   Can influence student outcomes

i. Identifying Essential Representations of Their Subject

The first dimension, the relationship between the teacher and their subject, is discussed more in the Curriculum Understanding Research Strand section of chapter 12, "Teachers Are Researchers." Deepening their understanding of this relationship must be a part of any teacher's professional-development journey.

The other four dimensions lie squarely in the realm where research from mind, brain, and education (MBE) science can help by suggesting strategies informed by research. This becomes clearer when we look at how Hattie subdivides these dimensions, where we see categories either right from our canon of research (such as "Expert teachers have high respect for students") or categories where the research suggests many strategies (such as "Expert teachers are proficient at creating an optimal classroom climate for learning").[2]

ii. Guiding Learning Through Classroom Interactions

Expert teachers are proficient at creating an optimal classroom climate for learning.
Expert teachers have a multidimensionally complex perception of classroom situations.
Expert teachers are more context-dependent and have high situation cognition.

iii. Monitoring Learning and Providing Feedback

Expert teachers are more adept at monitoring student problems and assessing their level of understanding and progress, and they provide much more relevant, useful feedback.

Expert teachers are more adept at developing and testing hypotheses about learning difficulties or instructional strategies.

Expert teachers are more automatic. ("Experts develop automaticity so as to free working memory to deal with other more complex characteristics of the situation, whereas experienced nonexperts do not optimize the opportunities gained from automaticity. These floaters are not incompetent but are not expert, as they do not use the advantages of the automaticity to put more back into the teaching act.")

iv. Attending to Affective Attributes

Expert teachers have high respect for students.

Expert teachers are passionate about teaching and learning.

v. Influencing Student Outcomes

Expert teachers engage students in learning and develop in their students self-regulation, involvement in mastery learning, enhanced self-efficacy, and self-esteem as learners.

Expert teachers provide appropriate challenging tasks and goals for students.

Expert teachers have positive influences on students' achievement.

Expert teachers enhance surface and deep learning.

If we can begin to identify traits of "expert teaching," it prompts a big question: how do we efficiently, and with appropriate speed, help create expert teachers? As a profession, education is not purposeful in doing this—it tends to hope it happens, often equating "experience" with "expertise." While sometimes this is the case, it is no guarantee.

Relying on time and hoping for the best also cannot be the most efficient or quickest way to create expert teachers. It is also, frankly, not a very professional way to run a profession. What if we could create a pathway, backed by research, to create expert teachers? What if we could create a blueprint for this professional pathway for schools to implement, and back this up with guidance to help them make this happen?

THE PROBLEM WITH CURRENT PROFESSIONAL DEVELOPMENT

For a profession that is centered on getting others to learn, much teacher professional development is bad to the level of comic irony. As mentioned in chapter 12, a review by the Teacher Development Trust found that only 1 percent of continuing professional development delivered to teachers was high quality. The research-tested evidence for exactly why what is being taught, is being taught, tends to be patchy or nonexistent. Many fit the "this worked in this school, you should try it, too" mold. These workshops flounder because it tends to be the package of small contextual details, all working together, that is the magic that made the method successful in that particular case.

These crucial packages of contextual details rarely ever get figured out or presented. Even if they were, the full contextual package necessary to replicate the "success" is likely to be fiendishly difficult to recreate. In addition, much professional development is based on principles that are just plain wrong. For example, teachers are still being taught that, according to Howard Gardner's *Theory of Multiple Intelligences*, students learn best when material is presented to them in their preferred learning style, even though this "neuromyth" has been debunked for many years.

This example highlights another fact. Gardner's original research publications are easy to find, very accessible to read and understand, and make very clear that this is not what he meant. This suggests an incredible intellectual laziness in those who create such professional-development "opportunities." Where is the evidence, where is the research to back up what we are asking teachers to learn? What other profession bases its continued learning and updating of practice on whims and hunches backed by no substantial evidence? To bring back this anecdote one final time, would you choose our mythical, much maligned heart surgeon who graduated from med school in the 1980s, but since then has relied on the few things he or she has happened to pick up along the way (some of which may be backed by research, but no one is really sure)?

Nevertheless, this all tends to work out okay, because professional development is usually presented in ways that, using our lens of MBE science, seem to be designed so that we remember as little of the experience as possible. It may be mediocre, but it is at least forgettable. It is hard to think of another profession that allows its continued professional growth and "staying up to date" to happen in such a slipshod way.

THE SOLUTION

We suggest that schools adopt a framework for professional development informed by MBE science. Such a framework would include the following five goals:

1. *Attract and retain the best and brightest into careers in education.*
   While this includes factors such as compensation and, perhaps, professional respect, it also includes professional satisfaction. Thus our framework must create pathways for challenging, interesting, inspiring professional growth in collaboration with like-minded colleagues.

   The pathway must allow teachers to see themselves at the cutting edge of a vibrant field, one rich with interdisciplinary collaborations that spread beyond their and others' classrooms into the world of research. The pathway must include opportunities to research, create, evaluate, reflect, and iterate. It must help teachers who are comfortable in their role in their own classroom, whether they are hankering for a bit more world-of-adventure or not, to see themselves in this role—an unusual one for our profession but one that is pretty common in other "professions."

2. *Practices should be informed by research from MBE science.* We use the word "informed" with purpose, and are grateful to Dr. Christina Hinton at Harvard's Graduate School of Education for this insight: that unlike phrases like "based on," the word "informed" leaves room for the art of teaching. It also hints at the fact that we often don't know a final definitive answer . . . and that is okay. This tends to be the nature of research. By making our practices be informed by research, we give ourselves this important role of trying something, evaluating how it works, discussing it with others and trying something a bit different, and thus we grow our personal practice and our group knowledge.

   MBE science suggests very lucrative gold mine claims to go prospecting in: an "unconscionable list" of practices that we probably should not be doing, as well as types of things that we should be doing. It is important to realize that MBE science only builds a useful scaffold *part way* to the branches of the delicious peach tree—the final meter is largely up to the teacher to figure out, for their particular discipline, for their particular age range, for their particular students given the other courses they are taking and the knowledge and skills they have. It also depends on the teacher's voice. Every teacher is unique; every teacher has a voice that is uniquely theirs. Each one should find and use it. This final meter is the challenge, but also the fun of it.

3. *Space should be made for a type of teacher research that promotes reflective practice for and amongst teachers.* MBE science tends not to offer a menu of "unpack, inflate, and use" strategies. The last step, the

final meter to the branches of the peach tree, the details of what best works, and how it best works, for *this* particular classroom, thankfully, is not written. And we say "thankfully" because if it were so prescriptive, it would probably be wrong in most scenarios. Furthermore, it would take the art and the fun out of teaching. It would suggest that our knowledge is static, that it has reached an apex of excellence, whereas we are just at a point where we have realized there are some fruitful areas to explore.

We need to have teacher researchers "playing" in these areas, trying things out, sharing their findings with other teachers and in a dialogue with the academic research world. We must create cultures of reflective practice—benefitting both the students in these teachers' classes and the profession we love as a whole. To make this happen, teachers need to be freed up from bureaucracies that too strictly dictate their daily practice. There are many pieces to the educational success of Finland, but it is hard to imagine how it could have happened without attracting smart, dedicated people to be teachers, training them well, and then giving them sufficient freedom, trust, and support (in addition to high standards and accountability) to lovingly craft the shape of their profession.

4. *Practice should be informed by evidence.* As Professor Rob Coe so eloquently drilled into us, *that strategy may well work, but what is your evidence?* We must train teachers and school leaders to collect quantitative and qualitative evidence, train them to analyze it, find the story that comes from it, and then use this to inform decisions on program, pedagogy, and professional development. That is, what teachers teach, how they teach it, and how they learn to get better and better at doing it. We must remember to look beyond numbers alone, what Carl Hendrick calls "the McNamara fallacy"[3] of education decision making.

5. *It must be high-quality professional development.* The delivery of and ongoing commitment to the professional-development framework must live up to the highest standards of pedagogy that it espouses. Following the principles of MBE science it itself lays out, and meeting the standards set forth in the Teacher Development Trust's excellent report, "Developing Great Teaching: Lessons from the International Reviews into Effective Professional Development"[4] are a good starting point.

THE PLAN: A PROFESSIONAL PATHWAY FOR TEACHERS INFORMED BY MIND, BRAIN, AND EDUCATION SCIENCE

The plan we propose includes four levels to be completed in order, as shown in figure 13.1 below. You will notice that this diagram includes a strand for school leaders, because we believe that the greatest positive change in teaching and learning will occur with these groups working in concert. It also recognizes the fact that some teachers will become school leaders. While the Center for Transformative Teaching and Learning (CTTL) offers programming to allow individual teachers to accomplish this, for this to be scalable we envision this as a guideline for schools or school systems to create their own model that fits their context.

Level 1: MBE Research Foundational Training, Novice
Level 2: MBE Research-Informed Teaching, Experienced
Level 3: MBE Research-Informed Teaching, Expert
Level 4: MBE-Informed Research Leader

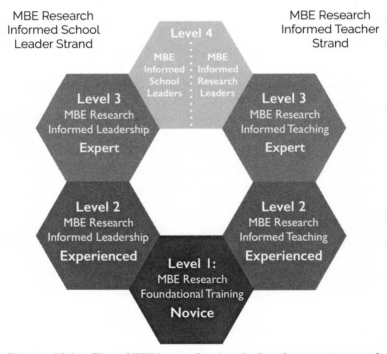

Figure 13.1. The CTTL's professional development growth framework: A professional pathway for teachers and school leaders informed by mind, brain, and education science.

Level 1: MBE Research Foundational Training, Novice

The goal of Level 1 is to create a knowledge base and, perhaps more importantly, a growth mindset in how teachers can be agents of change to improve learning for all students. This level also demystifies "educational neuroscience." It is our experience from having hundreds of teachers and school leaders from private, public, and parochial schools take these workshops that it shifts attitudes from "educational neuroscience is something intimidating that I cannot possibly do" to "it is something I can do, I must do, and here is how I can begin . . ." Furthermore, because the implications for learning in the classroom are so palpable, before the workshop is even finished it creates an appetite for "What can I do next? What is the next level of training?"

Level 1 covers four ideas, outlined below. It is the only level that is perhaps best introduced in workshop format—in part because it includes a lot of ideas that tend to be new to people, but also because it allows us time to build the transformational lens of the neuroeducational framework and guide teachers as they begin to apply it to their subjects, students, teaching, and assessing.

1. *Neuroscience 101: A student's learning brain*: Gain an understanding of the key parts of the brain and processes involved in learning. For example, how do we form memories? How does stress affect memory formation? What is myelination and how is it linked to learning? It is our belief that knowledge of how learning takes places influences teaching in ways that lead to better learning outcomes for students.

2. *Neuroplasticity for all*: Believe and embrace the concept of neuroplasticity. While there is a significant genetic component to intelligence (however we define it), there is also a significant environmental component. Students are constantly in the process of rewiring their brains—and how this happens will be influenced by the actions of the teacher. Furthermore, this is true for all students, the high fliers and "just fine" students as well as those with learning challenges.

3. *Equipping teachers with a neuroeducational framework*: We use a framework adapted from All Kinds of Mind that provides a common language and structure to describe all the various demands placed on a student's brain as they go through their school days. At our school, this framework is used as a lens by teachers to examine their subjects, their students, and how they teach and assess.

    What are the brain demands germane to my subject, and the ones my students are going to need to be good at to be successful in my subject in the future? What are the brain demands my teaching is going to challenge? What are the brain demands my assessments are going to challenge? How can I align these three so there is an essential fairness to the enterprise of learning? What brain demands have I placed on my students recently? What brain demands do I know I am going to place on them soon? In what brain demands are students in my class currently strong and weak? How can I balance things out or mix things up so that the same kids do not find it easy all the time or hard all the time?

When we say this is a lens for teachers to examine their students, we mean each individual student, identifying areas of strength and challenge, and having them identify and reflect on areas of strength and challenge. It is also a lens to help teachers observe, analyze, and intervene when students are finding work too tough or too easy. As part of their training, teachers use the neuroeducational framework to analyze students, assignments, and curriculum, and to suggest possible courses of action to take. One additional and very important outcome is that teachers learn a common vocabulary, so are able to discuss multiple aspects of their work with this common neuroeducational lens.

4. *Teaching and learning strategies*: There are a number of strategies for teaching and learning stemming particularly from "Neuroscience 101" and "Neuoplasticity for all." Some are information teachers should know, some are things teachers should either do or not do.

Level 1 MBE-Informed Strategies: The Implications of Neuroscience 101 and the Concept of Brain Plasticity
Information to Know

- Are you aware that your students' brains remain plastic throughout the time you will teach them and beyond, that while there is a significant genetic component to intelligence, there is also a significant environmental component, which means that how you choose to teach has an effect on how their brains rewire themselves?
- Are you aware that while students all have individual differences in brain demands at which they are stronger and weaker, labeling students as, for instance, "auditory learners" actually hurts their ability as students? Are you aware that a better plan is for each student to become aware of his or her personal strengths and weaknesses, and

what strategies he or she can use to help them become, because of the brain's plasticity, a better and better student?

- Emotion is inseparable from learning in the brain. Students need to recognize the impact stress, fear, and fatigue have on their brain's ability to learn, and what they can do to maximize the use of the higher-order thinking, executive functioning, and memory parts of their brain. Do you believe that part of your role as a teacher is to share strategies with your students to help them constructively cope with emotions?
- Are you aware that stress impedes learning (and, by contrast, a positive emotional environment increases attention and learning)?
- Are you aware that emerging research on mirror neurons suggests that "emotions are contagious," that your demeanor affects your students' and colleagues' demeanor?
- Research supports that when teachers understand principles from educational neuroscience, it changes how they teach, how they assess, and improves learning for their students. Do you believe that the more you know about how the brain learns, the more you will change your practice in ways that will increase every student's performance?

General Things to Do

- What does your classroom look like? Is it stimulating but not over-stimulating? Does it change? Do you show recent student work that changes to keep up to date? Do you create novelty? Are the key objectives of your class prominently displayed in the room?
- Do you praise effort rather than achievement?
- Students need to know that "effort matters most" and the brain has the ability to rewire itself (plasticity). Do you coach your students that deliberate effort can rewire the brain and lead to enhanced academic achievement?
- Do you stress to your students the importance of sleep for memory consolidation?
- Do you know the individual learning strengths and weaknesses of your students? Do you evaluate each student's learning strengths and weaknesses through direct observations rather than just using labels (such as "unmotivated," "lazy," "bright")?
- Do you have high expectations of all your students, irrespective of their individual learning strengths and weaknesses?

Specific Things to Do in Class

- Do you provide students with specific learning objectives?
- Do you use brain breaks or movement during your class to reset attention?

Level 2: MBE Research-Informed Teaching, Experienced

Things to Do, Things Not to Do

Following the empowerment that we have found Level 1 tends to bring, teachers often ask, "So what should I be doing in my class?" Research from the fields of MBE science suggests "the unconscionable list," practices that teachers should not be doing, and also ones that they should be trying. The goal of the next two levels is to inform teachers about both, to present them with just the right amount of research evidence behind the thinking, and to give them accessible gateways into the academic research for them to explore more if they wish.

To make this task more digestible, it has been split into two. Level 2 focuses on strategies that can be readily implemented in a wide range of classrooms—the teacher just has to figure out how it is going to work in their context. Many are low-hanging fruit—strategies, supported by solid research, that can make a significant difference. Level 3 focuses on strategies that are more connected to how a teacher teaches that particular subject, and where successful implementation is likely to be more involved. In both levels, some of the strategies are teacher-centric—requiring teachers to change their behavior to include more MBE-informed strategies—while some are student-centric—requiring students to adapt behavior.

As mentioned earlier, the research provides scaffolding to get close to the fruit-bearing branches of the peach tree, but the last meter is often down to the teacher to figure out, as the effectiveness of any strategy usually rests on the wide spectrum of contextual details of any particular case—the subject, the age of the students, the range of abilities, what prior knowledge they have, for example, but you can think of many more. Teachers must tease out how the strategies research suggests are relevant to their own context. In the words of Carl Hendrick, Head of Research at Wellington College:

> Currently in education research there is the constant refrain, "What works?" A better phrase for me would be "what works in your context?" . . . Instead of passively accepting the "stone tablets" of research we should be engaged in a constant dialogue with research, questioning it, challenging dogmatism,

teasing out relevance to our own context and our own individual problems in a sort of "detached attachment." We should be constantly reviewing our own preconceptions and refining our practice through this process of oscillation and reflection. We should reconcile ourselves with the irreconcilable nature of the classroom. What may work on Tuesday period 3, might be a disaster on Thursday period 2 with the same class and the same teacher for a variety of reasons, some that we may "know" but many that are simply unknown.

Teacher Research

"Know thy impact," as Hattie reminds us. Are the strategies you are trying increasing learning? How do you go about measuring this anyway? This is why Level 2 and 3 both include teacher-researcher protocols—Level 2 is designed to be very accessible and implementable by a busy teacher; Level 3 is designed to be a more robust research methodology, adapted to make it doable by a busy teacher, but it is unashamedly more involved. Common to both is the aim of promoting reflective iterative practice that is informed by research and, hopefully, collaborative.

The ideal scenario is for Level 2 and 3 to be supported by a research lead—preferably within the school. But our framework is designed to generate capacity for mentorship as it goes. The goal of Level 4 is to create research leads; the goal of Level 3 is to create mentors for those in Level 2. The key, then, is how to break into this chicken-and-egg scenario. We must and will offer training to create and support the first generation of research leads in a school.

Level 2 includes the following five steps:

i.   Self Assessment

- What practices informed by MBE science am I currently *aware* of? (I may be doing them, but I may not—this is an assessment of knowledge, not practice.)
- What am I currently *doing* that is a practice informed by MBE research (I may already be doing things that MBE science research supports without knowing it—this is an assessment of practice not knowledge.)
- What practices or ideas that I believe in are actually neuromyths?

ii. Teaching and Learning Strategies

- What practices does research suggest I NOT do?
- What is the research behind these suggestions?

- What strategies does research suggest I try to implement? What is the research behind these?
- What might these strategies actually look like in the context of my particular classroom?

   iii. Research

- What are some gateways into the research so I can go exploring?
- What skills do I need to become an explorer of published research?
- Teacher research protocol: what skills do I need to try something out, document, evaluate, and refine the strategy? This protocol is outlined in chapter 12, "Teachers Are Researchers."

   iv. Mentoring

- Receiving ongoing support to help implement strategies and conduct teacher research to evaluate the impact of strategies
- Receiving ongoing support to help navigate MBE research literature

   v. Community

- Collegiality—because, depending on the school, the world of an MBE-informed teacher can be, initially at least, a lonely one
- Opportunities to disseminate ideas and findings
- Opportunities to learn from or be inspired by others

Level 2 MBE-Informed Strategies
Things NOT to Do
See chapter 3 for this list of neuromyths.

Things to Do: Twelve Areas Backed by Solid Research That Should Impact Your Students' Learning

For most, possible starting questions are given—use these as a guide and not definitive boundaries of scope. Conduct your own deeper research on these ideas. Work collaboratively and iteratively to figure out ways to implement them in your context. Attempt to use the simple teacher-researcher protocol outlined in chapter 12 to "know thy impact."

   i.   Using feedback to improve student learning

- When you return work, do you provide scaffold feedback, allowing students time to struggle to correct their errors, rather than simply marking it right or wrong or giving them the answer?

- Do you give students opportunities to redo or edit work rather than make everything one-and-done?
- Do you return work quickly?

ii. Using formative assessments to improve student learning

- Do you provide frequent, low-stakes formative assessments throughout the learning process for students to practice recalling?

iii. Using metacognition to improve student learning

- Do you help your students think about their own learning more explicitly? For example, what strategies did I use? How well did they work? What could I have done differently?
- Do you help students pause to break higher-order thinking tasks into recognizable steps?
- Do you provide scaffolding to help individual students develop strategies for learning that work for them? (including broad skills such as memorizing, organizing, and the logistical management tasks of being a student, as well as for specific skills and learning tasks in your class).
- Do you help students recognize their individual strengths and weaknesses as learners, help them develop strategies to aid each, and keep them cognizant of the implications of neuroplasticity, which means that their hard, smart effort will help rewire their brains to make them stronger students?

iv. Teaching memorizing strategies, such as active retrieval methods, to improve learning

- Do you encourage students to self-test or use active retrieval methods to study rather than just rereading class notes or their textbook?
- Do you coach students to space their studying rather than use massed studying?

v. Using reflection to improve student learning

- Before each of your assessments, do you have students reflect on how they will study and what demands a particular assessment will have on their brains?

- At the end of each of your assessments, do you require students to answer a few reflection questions concerning how they studied, how effective their methods were, or what they think their grade might be?
- Do you use exit tickets in the last few minutes of class, requiring students to recall information from the class and to pose a question such as, "What question do you still have after class today?"
- Do you use journaling or portfolio assessment as ways to incorporate student reflection?

vi.   Using student choice to improve learning

- Do you ever give students any choice in what they are studying in order to increase engagement?

vii. Building a growth mindset through the use of specific learning strategies and reflection on how they are working

- Do you give constructive criticism and specific feedback about effort and strategies to help students develop their own personal, effort-based learning goals and strategies?

viii. Using peer tutoring to improve student learning

- Do you make use of peer tutoring?

ix.   Providing students with explicit instruction on how to plan and organize in order to help them enhance their executive functioning skillset

- Do you provide students with explicit instruction on how to plan and organize in order to help them enhance their executive functioning skillset?

x.   Providing students with experiences and environments that aid identity validation, and eliminating factors that may cause or enhance identity threat

- Do your actively work to ensure that your students feel heard, listened to, and known?
- Do you actively work both to provide students with experiences and environments that aid identity validation, and to eliminate factors that may cause or enhance identity threat?

xi.   Integrating the visual and performing arts into non-arts subjects to create knowledge transfer that improves learning

xii. Using play constructively to increase student engagement and improve learning

Final Word: Content

- Some of the strategies described above may save time; for example, they might increase attention, or enable students to be more independent learners. Others, however, will take time. Are you prepared to lessen the amount of content you teach in order to provide more meaningful learning experiences that have a greater chance of leading to deeper learning, increased storage in long-term memory, and the acquisition of new skills?

At the End of Level 2, Teachers Will Be Able to Do the Following:

1. Have stopped doing practices that MBE research suggests should be stopped.
2. Be aware of strategies informed by MBE research.
3. Implement strategies in the context of their particular classroom.
4. Evaluate the effectiveness of strategies implemented using a defined but flexible teacher-research process.
5. Be connected to a support network of other MBE-informed teachers.
6. Disseminate their findings within this network (and beyond).

Level 3: MBE Research-Informed Teaching, Expert

Whereas Level 2 focuses on general classroom strategies, Level 3 focuses on strategies that are deeply connected to how teachers teach their particular subject. These strategies are often highly focused, used to improve the learning of a particular idea, skill, or content area. They are even more dependent on context than Level 2. As before, some of the strategies are teacher-centric while others are student-centric. The goal of Level 3 is to create exceptional implementers of MBE-informed strategies and mentors for teachers in Level 2.

In addition, we would like teachers to become more familiar with a more robust and rigorous research methodology. As mentioned in chapter 12, the details of this process are beyond the scope of this book, but a starting point for this is the excellent work in this area by Dr. Stuart Kime at Durham University in the United Kingdom.

Level 3 follows the same five steps as Level 2: (1) Self Assessment, (2) Teaching and Learning Strategies, (3) Research, (4) Mentoring, and (5) Community. However, there is an additional emphasis on mentoring other teachers in Level 2.

Level 3 MBE-Informed Strategies: A Greater Focus on Subject and Content-Specific Strategies
General Things to Do
Connectedness and Stress

- Do you actively work both to provide students with experiences and environments that aid identity validation, and to eliminate factors that may cause or enhance identity threat?

Brain Plasticity

- Students need to know the anatomy of their brains, especially the roles of the prefrontal cortex, amygdala, and hippocampus in learning. Do you teach your students about this within the context of your course?

Specific Things to Do in Class
Connectedness and Stress

- Is your curriculum designed so that students will see meaning, relevancy, or emotional connections to their own lives?
- Do you actively work at helping students see the meaning and relevancy, or make emotional connections between the work they are doing in class and their own lives?

Class Design

- Do you offer students choice in subject matter and/or means of assessment in order to increase engagement?
- Do you apply the primacy-recency effect to how you design your classroom lessons? That is, class periods should be designed with an understanding that what students will recall most is what takes place in the first part of the class and what students will recall second most will take place in the closing minutes of class.

- Do you use play constructively at all grade levels to help build student engagement?

Memory

- Do you use interleaving when you design your curriculum in order to enhance long-term memory storage?
- Do you stress the importance of sleep for memory consolidation? And do you give high-quality, well-judged homework to help make this possible?
- Do you coach students on managing their time to help this happen?

Multisensory/Multimodality Teaching

- Do you routinely teach in a variety of modalities? Do you attempt to vary modality to better suit the content?
- Do you assess in multiple modalities?

Arts Integration and Visual Thinking

- Students need opportunities to transfer their knowledge through the visual and performing arts. Do you integrate art into the core content of non-arts subjects to enhance learning?

Reflection and Metacognition

- Students need opportunities to reflect and think metacognitively on their learning strategies and performance. Do you build in moments for metacognition and reflection at appropriate moments during your class?

Final Word Again: Content

Some of the strategies described above may save time; for example, they might increase attention, or enable students to be more independent learners. Others, however, will take time. Are you prepared to lessen the amount of content you teach in order to provide more meaningful learning experiences that have a greater chance of leading to deeper learning, increased storage in long-term memory, and the acquisition of new skills?

At the End of Level 3, Teachers Will Be Able to Do the Following:

1. Be aware of and implement a broad repertoire of strategies informed by MBE research.
2. Evaluate the effectiveness of implemented strategies using a more rigorous teacher-research process.
3. Be a mentor for Level 2 teachers in their school, working to implement and evaluate strategies in their classroom.
4. Disseminate their findings.
5. Communicate stories of MBE-informed teaching in action—what teachers are doing to improve learning and why—to a wider audience using their school's communication network.

Level 4: MBE-Informed Research Leader

Schools need research champions, carrying the flame of evidence-informed, reflective, and iterative practice in their schools. They need people who understand the personal satisfaction, and better teaching and learning, that come from engaging in research work. They need people to help demystify the process of research, and to give other teachers courage, knowledge, and resources to help them do it. They need people to help plan what to study and how to study it. They need help in analyzing their results and disseminating their findings.

Schools need people to tell the stories of how research is informing practices that lead to better learning. Schools also need to help bring some degree of cohesion to the various tendrils of research that might be happening. Schools need research leads, and research leads to be informed by evidence from the field of MBE science.

Research is a skill that, for most teachers, must be acquired. Therefore, training must be given so that Level 4 teacher "research leaders" can effectively mentor others in the following:

How to find and evaluate MBE research literature.
How to plan a teacher research study, and what the limitations of such a study might be.
How to conduct research studies.
How to mentor and support teachers involved in research studies.
How to collect evidence and data from different stakeholders—what data are valuable and how might I get them?

How to evaluate the findings of a research study.
How to define the limits of what the study does say and does not say.
How to use the results of a study iteratively to plan future work.
How to disseminate effectively the results of teacher research studies.

Level 4 teachers have a leadership role to play, so training must also be given in the following:

How to be an effective mentor.
How to manage a number of research studies going on at once.
How to persuade other teachers to take that leap of faith and begin their own journey along this MBE-informed professional-development pathway.

This skillset and knowledge creates an interesting and unfamiliar type of leadership position in a school, and we argue it is a compelling one. It is a leadership role for teachers who want to dedicate themselves to being excellent at their profession, to be experts, and who want to help their colleagues along this pathway of professional growth, too.

We expect that Level 4 teachers will continue to try implementing new MBE-informed strategies, conduct research on the effectiveness of what they are trying, and disseminate what they find. In addition, we anticipate that they will lead, support, and promote the teacher research work of others—being a catalyst for a wave of creative, collaborative energy centered on professional growth that aligns with the school's mission.

At the End of Level 4, Teachers Will Be Expected to Do the Following:

Have a strong knowledge of the MBE-informed strategies in Levels 1, 2, and 3
Have a strong knowledge of MBE-informed "things not to do"
Be able to implement MBE-informed strategies in the context of their own classes
Help others implement MBE strategies in the context of their particular class
Conduct teacher research projects
Help others conduct teacher research projects
Share their knowledge through workshops or publications to expand the use of MBE-informed strategies in their field

It is important that school leaders recognize the "value added" of having teachers who have chosen to follow this professional path. Remember, expert teachers are different from experienced teachers. It is expert teachers who make the greatest difference in learning. Experience does not automatically create "expert."

We can get to "expert" quicker if we are mindful and plan rather than leaving it to chance. We now know the kind of practices that expert teachers should and should not be doing. It will vary from subject to subject and grade to grade; it will vary based on the contexts that each school has; it will vary based on each teacher's own unique voice. But, nevertheless, we can begin to codify the sorts of practices that expert teachers will do and will not do. So let's create pathways to develop expert teachers. And let's do so in a way that has a built-in capacity for creating mentors and leaders along the way so that schools have a chance of growing this change.

In doing so, we may also create a pathway for increasing professional satisfaction: a pathway for teachers seeing themselves at the cutting edge of a dynamic field; feeling as if they are equipped with tools to "make a difference"; helping other teachers with their professional growth; having a significant degree of control over their own professional development, and in a way that allows them to work with others.

If we want to be bolder, we can create ways to incentivize financially the four steps on our professional pathway, and it should not be prohibitively hard to devise systems of accountability for doing so.

We hope that a professional pathway to being an expert teacher will align with the core mission of most schools—for excellent teaching must surely be one of the elements at the core of a school's mission. If it is not, this is where we must begin. We can equip teachers with knowledge and training. We can offer support and plan mentorship for their journey. But each individual teacher needs time and intellectual freedom to flourish, and in an environment where this feels valued. Maybe this is the greatest challenge. Is there a will to create the environments that will let it happen?

In the words of Toni Morrison, "As you enter positions of trust and power, dream a little before you think." We have formulated our research-informed plan. It is not such a crazy plan. When you think about it, it is similar to how we train doctors. We have not deliberately made it this way, but when you follow the steps of how current research can and must be used to influence practice in a field, this is where you end up. It is, put simply, a professional way to conduct a profession.

Think back to your school days and picture the teacher you had who made you most wonder, "How on Earth is this person still teaching?"—

a teacher who did the same old things year after year without too much thought. Imagine the training this teacher might have had from the moment he or she left college to the day he or she encountered you. Now switch locations and imagine you are being wheeled into an operating theater for heart surgery. Look around the room at all the different medical professionals gathered there for the operation. Now imagine if any one of them—not just the surgeon, any one of them—had received the same type of ongoing professional development training that your teacher had.

Without looking back from this page, what are the *three* most salient points you take away from this chapter of *Neuroteach*?

What are *two* things you would like to do "tomorrow" with the information you learned from reading this chapter?

What is *one* question you have after reading this chapter?

NOTES

1. Emphasis added. John Hattie, *Teachers Make a Difference: What Is the Research Evidence?* Australian Council for Educational Research, Melbourne, 2003, http://www.decd.sa.gov.au/limestonecoast/files/pages/new%20page/PLC/teachers_make_a_difference.pdf.

2. Hattie, *Teachers Make a Difference.*

3. "The McNamara Fallacy and the Problem with Numbers in Education," *Chronotope*, http://chronotopeblog.com/2015/04/04/the-mcnamara-fallacy-and-the-problem-with-numbers-in-education/ (accessed May 7, 2015).

4. P. Cordingley, S. Higgins, T. Greany, N. Buckler, D. Coles-Jordan, B. Crisp, L. Saunders, and R. Coe, *Developing Great Teaching: Lessons from the International Reviews into Effective Professional Development* (London: Teacher Development Trust, 2015).

CONCLUSION

The 10 Percent Challenge

Have a bias toward action—let's see something happen now. You can break that big plan into small steps and take the first step right away.

—Indira Gandhi

What you can do, or dream you can do, begin it;
Boldness has genius, power and magic in it.

—John Anster's translation of *Faust*

Kismet. Two things happened in the twelve hours before we sat down to write this conclusion. The first was an exchange on Twitter:

If there's teaching going on, but the students aren't learning, is it really teaching?[1]

To which the posted reply was:

Interestingly the next question is how you would know if learning was really going on (as opposed to performance)?[2]

These two quotations taken together go a long way to capture what we are trying to do through *Neuroteach*. How do we know that any teaching and learning are going on in the first place? And then, how can we use a diversity of research to inform what we do to make the teaching and learning of each individual teacher, and the outcomes for each individual student, as good as they can be?

The second moment came as we sat down with a collaborator, Dr. Sheila Walker from Johns Hopkins Graduate School of Education and Bloomberg School of Public Health, to discuss a proposal we had received for a possible online mind, brain, and education (MBE) course. In one of those serendipitous moments of insight, Dr. Walker said that the first video for the course should begin by describing how this series of videos will not actually be able to teach you all that you need to do. It would be a start, and starting is, of course, important, and often treacherously difficult. But how will you continue? MBE research-informed strategies for teaching and learning will get you so far, but you have to run the last leg of this relay race yourself. This is a fundamental part of being a research-informed teacher.

The research, this book, those videos—all will suggest strategies that might be made to work to enhance learning for your students. But you get to figure out exactly how to make them work in your context, your classroom, your kids, your school, and your families. This is both the challenge and the fun of it. Collaboration will help, as will mentorship. But our course videos could not hope to tell you everything. The videos would help you define some awesome sandboxes to play in, the right sandboxes we feel, but you have to step into them and play.

But none of this will work without a mindset shift that is fundamental for the professionalization of the profession of teaching: reflective, iterative, research-informed practice must be how we proceed. However good I am as a teacher, I will be better next year—and I stand a good chance of doing so because *these* are the research-informed avenues I will be exploring. In fact, I know I am a better teacher this year than I was last year because *this* is what I did, *here* is the research that informed what I did, and *this* is how I measured its effectiveness.

Our (at the point of writing, hypothetical on-line course) videos will help you start down this path, and more will support you along your journey, but you need to commit to the journey of being an MBE-informed teacher. Your journey is and will be different than any other teacher because your voice as a teacher and the context of your classroom are different than any other. You may take the same research as someone else and end up in a different place—and that is okay. In fact, it is more than okay, it is exactly what's needed. The similarity is that both will arrive by a process of reflective, iterative, research-informed (and hopefully collaborative) practice. This, perhaps, is the most important message of our book—be on an evolving-evidence, research-informed teaching journey.

Our sensitivity to this point comes from our experience giving workshops at a wide variety of schools. We sense that there is a tendency in a few places to say, "That was a great workshop, thanks, I get it now." The participants then tick a box to say, "Right, we have done mind, brain, and education now" before moving on to the next professional-development temptation, like a magpie with attention challenges in a costume jewelry store. They are missing the point.

The potential for great change, for great improvements in teaching and learning that will benefit *all* students, comes from the cumulative effect of reflective, iterative, research-informed practice, performed by a number of people over the course of a number of years. This is not a one-and-done "event," it is a shift in how you are going to do school. With a critical mass of teachers and school leaders, it can bring a creative energy around great teaching to your whole school, an energy that the students mirror and thrive on. Remember, an expert teacher is different from an experienced teacher, one does not automatically beget the other, and the road to being an expert is one that any teacher at any point in their career can purposefully take. Every one of our students deserves an expert teacher.

Change will happen over time as you commit to it. This is true both individually and institutionally. It is why we suggest you take up our "10 Percent Challenge." We challenge you, each and every year, to change 10 percent of what you do in ways that are informed by MBE research. What things will you do?

We must make one admission, however. In a book based on research, there is a final dose of irony in how we reached this number: 20 percent is too big and anything ending in a 5 percent looks annoying. But 10 percent? Undoubtedly, 10 percent is a challenge, but a manageable one, not too scary. The cumulative effect of even a few years of 10 percent is massive. Even if some of your changes "cancel out" because they end up warranting further changes, where will you be in five years' time? More importantly, where will your students be? And how do you think they will react to having, standing before them, a teacher who is modeling research and reflection, using evidence to inform what he or she does, and working hard and smart toward mastery goals?

How can schools support this change? We strongly believe that a common framework and language are crucial to underpin system-wide implementation of a new initiative—and this is what we begin to present in chapter 12. Another vital element for successful school-wide change is the presence of effective mentorship—so the framework we presented in chapter 12 was designed to generate leveled mentorship capacity as it grows.

The vision to cultivate a research lead in a school is a powerful step forward, and perhaps a game-shifting statement to the school community. A point person committed to helping a school be "research informed" helps, unsurprisingly, the school *be* research informed. That person is a fountain of knowledge, plans and assists in teacher research, and connects the school to the greater research world. But he or she is also crucial in connecting people, helping the work that teachers undertake align with the school's strategic mission, and stopping ideas and opportunities from falling through the cracks.

Being a research-informed school is further aided by leaders who recognize that time and support are needed for the faculty to cultivate a culture of iterative, reflective, collaborative practice. While this list is not exhaustive, the point we are trying to make is that becoming a school whose teaching and learning are informed by MBE research is something that *can* be planned.

The New Teacher Project's report titled "The Mirage: Confronting the Hard Truth About Our Quest for Teacher Development"[3] received significant press when published in 2015, outlining as it does what at first glance seems like a grim picture of a lot of time and money spent on teacher professional development with scant improvements to show for it. But we think this is actually a glimmer of optimism—it suggests that necessary time and money for positive change do actually exist, but we are just not using them effectively. We put it to you that MBE research can inform both what we need to do and how we go about doing it.

As we draw our book to a close, we draw your attention back to the dedication two hundred pages ago. Every student deserves a teacher who understands how the brain learns best—what we know about how learning occurs, which strategies are effective and which are not, and which myths we need to fight against. Settling for anything less would be unconscionable.

NOTES

1. From Paul Bambrick-Santoyo (@paul_bambrick) via Carl Hendrick (@C_Hendrick).
2. From David Weston (@informed_edu).
3. "The Mirage | TNTP," http://tntp.org/publications/view/the-mirage-confronting-the-truth-about-our-quest-for-teacher-development (accessed October 1, 2015).

APPENDIX I

Readings, Research, and Resources

The following is a list of books, articles, and websites that have been integral to our evolutions as neuroteachers. This is just a start, and we encourage you to go to http://www.thecttl.org/neuroteach for a more complete and evolving list, and we welcome your suggestions for new resources to share with the neuroteach professional learning community.

BOOKS

Peter C. Brown, Henry L. Roediger III, and Mark A. McDaniel, *Make It Stick: The Science of Successful Learning*

Benedict Carey, *How We Learn: The Surprising Truth about When, Where, and Why It Happens*

Carol Dweck, *Mindset: The New Psychology of Success*

David Didau, *What If Everything You Knew about Education Was Wrong?*

Kurt W. Fischer and Mary Immordion-Yang, *The Jossey-Bass Reader on the Brain and Learning*

Howard Gardner, *The Unschooled Mind: How Children Think and How Schools Should Teach*

Susan Greenfield, *Mind Change: How Digital Technologies Are Leaving Their Mark on Our Brain*

Mariale Hardiman, *The Brain-Targeted Teaching Model for 21st-Century Schools*

John Hattie, *Visible Learning: A Synthesis of over 800 Meta-Analyses Relating to Achievement*

John Hattie, *Visible Learning for Teachers: Maximizing Impact on Learning*

Mary Helen Immordino-Yang, *Emotions, Learning, and the Brain: Exploring the Educational Implications of Affective Neuroscience*

Robert Janke and Bruce Cooper, *Errors in Evidence-Based Decision Making: Improving and Applying Research Literacy*

Paul Howard Jones, *Introducing Neuroeducational Research: Neuroscience, Education and the Brain from Contexts to Practice*

Eric R. Kandel, *In Search of Memory: The Emergence of a New Science of Mind*

Ron Ritchhart, *Creating Cultures of Thinking: The 8 Forces We Must Master to Truly Transform Our Schools*

Ron Ritchhart, Mark Church, and Karin Morrison, *Making Thinking Visible: How to Promote Engagement, Understanding, and Independence for All Learners*

Martin Robinson, *Trivium 21st Century (Preparing Young People for the Future with Lessons from the Past)*

Vanessa Rodriguez, *The Teaching Brain: An Evolutionary Trait at the Heart of Education*

Hana Ros and Matteo Farinella, *Neurocomic*

Todd Rose, *The End of Average: How We Succeed in a World That Values Sameness*

Jack P. Shonkoff and Deborah A. Phillips, *From Neurons to Neighborhoods: The Science of Early Childhood Development*

David A. Sousa, ed., *Mind, Brain, and Education: Neuroscience Implications for the Classroom*

Tracey Tokuhama-Espinosa, *Making Classrooms Better: 50 Practical Applications of Mind, Brain, and Education Science*

Daniel Willingham, *Why Don't Students Like School? A Cognitive Scientist Answers Questions about How the Mind Works and What It Means for the Classroom*

Daniel Willingham, *When Can You Trust the Experts? How to Tell Good Science from Bad in Education*

Judy Willis, *Research-Based Strategies to Ignite Student Learning: Insights from a Neurologist and Classroom Teacher*

ONLINE RESOURCES

"Applying Science of Learning: Infusing Psychological Science in the Curriculum," editors Victor Benassi, Catherine Overson, and Christopher Hakala, from the American Psychological Association

The Dana Foundation: Your Gateway to Responsible Information About the Brain

"Developing Great Teaching," from the Teacher Development Trust, http://tdtrust.org/about/dgt

"Improving Education: A Triumph of Hope Over Experience" by Rob Coe, Centre for Evaluation and Monitoring

"Mind, Brain, and Education: The Student at the Center Series" by Christina Hinton, Kurt Fischer, and Catherine Glennon, *Mind, Brain, and Education*, March 2012

The "Teaching and Learning Toolkit," from The Education Endowment Foundation (EEF)

"Understanding the Brain: The Birth of a Learning Science" (OECD)

"What Makes Great Teaching?" by Robert Coe, Cesare Aloisi, Steve Higgins, and Lee Elliot Major, from the Sutton Trust

PROFESSIONAL DEVELOPMENT FOR TEACHERS, STUDENTS, AND SCHOOL LEADERS

Brain and Learning (online course designed by NEA Foundation and Harvard GSE, http://www.neafoundation.org/pages/courses)

Brain Awareness Week, The Dana Foundation (http://www.dana.org/BAW)

Brain Targeted Teaching Model (http://www.braintargetedteaching.org)

Mind, Brain, and Education at Harvard Graduate School of Education (http://www.gse.harvard.edu/masters/mbe)

Mind, Brain, and Education (MBE) Research Engagement Professional Growth Framework (http://mberesearchengagementframework.org)

Mind, Brain, and Teaching Program at Johns Hopkins School of Education (http://education.jhu.edu/Academics/certificates/mindbrain)

Neuroscience and the Classroom (Annenberg Foundation free online course, http://www.learner.org/courses/neuroscience)

Research Schools International (http://rsi.gse.harvard.edu)

Teaching All Kinds of Minds (http://www.allkindsofminds.org)

The Center for Transformative Teaching and Learning, Resources and Blog (http://www.thecttl.org)

APPENDIX II

Self-Reflection Tool

A PERSONAL MBE SCIENCE RESEARCH-INFORMED STRATEGIES CHECKLIST

In many parts of *Neuroteach*, we discussed how metacognition and reflective practice are critical to deepening learning for students. These habits of mind are equally important for teachers and school leaders. Use this checklist to evaluate how well you are *currently* applying some of the following research-informed strategies, selected from each chapter, to the design of your class, program, or work with each individual student. And yes, this is also an example of the spacing effect—we hope this exercise helps stick these important ideas in your memory. Use the following scoring scale:

  5: very comfortable with this idea
  4: somewhat comfortable with this idea
  3: neither comfortable nor uncomfortable
  2: somewhat uncomfortable with this idea
  1: very uncomfortable with this idea

Or, if you prefer, and with the view that choice and novelty can also be beneficial, use this five-point smiley face scale:

Chapter 1: Educational Neuroscience for All

○ i. Teaching is both an art and a science. There is evidence that should be used to inform practice—but exactly how these strategies work in the context of your classroom, your students, and your school need to be figured out by you.

○ ii. Implementing strategies informed by mind, brain, and education (MBE) science works for all students: the most advanced student, the often overlooked "just-fine-student," *and* the struggling student.

○ iii. Executive functioning is actually an important set of skills that schools can influence the development of: problem-solving, prioritizing, thinking ahead, self-evaluation, long-term planning, calibration of risk and reward, and regulation of emotion. *All* students can benefit from carefully scaffolded practice that helps them get even better at this crucial set of school, job, and life skills.

○ iv. Knowledge of a neurodevelopmental framework allows teachers to align the brain demands inherent in the material they want to teach with the brain demands of how they teach it and the brain demands of how they assess it.

○ v. Although each student has individual learning preferences, all students learn best when taught in a variety of modalities. The best modalities to use will vary from concept to concept. Teachers should differentiate based on content, not learning style.

Chapter 2: A Formative Assessment

○ i. The more teachers understand principles from educational neuroscience, the more they will believe in a student's ability to improve their academic performance.

○ ii. Brains are not able to multitask—there is a transaction cost for trying to get your brain to do so. Endeavor to teach your students to appreciate this—it will make them more efficient learners.

○ iii. Human brains seek and often quickly detect novelty. Teachers can use well-chosen moments of novelty to increase engagement and motivation.

○ iv. Integrating the arts into the curriculum enhances learning and understanding.

Chapter 3: The Top Twelve Research-Informed Strategies Every Teacher Should Be Doing with Every Student

○   i.   Lecturing is a good teaching method at the right time for content delivery, and a skill students should develop. But be careful how and when you use it, and don't let it dominate.

○   ii.   Likewise, tests are good and have their place, but assessment should not be dominated by tests, particularly multiple-choice tests.

○   iii.   Avoid applying simple labels to students, such as "lazy" or "smart." Instead, make judgments based on observations, and continually work to review and revise your thinking based on further observations.

○   iv.   Do not define students by an individual style, such as "this person is an auditory learner, that person is a kinesthetic learner," as these labels are simplistic and ultimately misleading.

○   v.   Students need frequent opportunities during the school day to play.

Chapter 4: How Much Do We Need to Know about the Brain?

○   i.   While there is a significant genetic component to the architecture of each individual's brain, the environments and experiences which we expose our brains to cause them to change, and continue to do so throughout our lives.

○   ii.   By deliberate effort—working harder and smarter—we can influence the type of changes that take place. This is the concept of neuroplasticity.

○   iii.   Common neuromyths include the assertions that we only use 10 percent of our brains, people are either dominantly left brained or right brained, that brain development is essentially finished by secondary school, and that individuals learn better when they receive information in their preferred learning style.

○   iv.   Cognitive and emotional areas are integrated in the brain, so that emotion is necessarily integrated with learning.

Chapter 5: A Mindset for the Future of Teaching and Learning

◯   i.   Having a growth mindset requires the development of clearly defined strategies for improvement and the enlistment of support, advice, and guidance from others. Simply praising effort rather than achievement, while important, is not enough.

◯   ii.   Showing students how the brain is wired and can be changed through "effortful learning" enhances their belief in their ability to change their brain.

◯   iii.   Success takes practice, lots of practice, lots of *deliberate* practice: working on technique, seeking critical feedback, and focusing ruthlessly on shoring up weaknesses.

◯   iv.   There is a direct relationship between hours of practice and achievement—the idea of "naturally talented" people who can take a shortcut to mastery without putting in the hours of practice is a myth not backed up by evidence. There are no shortcuts.

◯   v.   Teachers should provide appropriate scaffolding to help students recognize what high-quality deliberate practice means for them in the context of their own individual learning strengths and weaknesses, and the context of the academic tasks they are being set.

◯   vi.   Teachers should help students develop an iterative process of trying strategies, evaluating how well they work, then refining them or finding new ones.

◯   vii.   We need to challenge students. Our brains are wired to learn; it is fundamental to being human.

◯   viii.   Cultivating an ability to deal positively with failure, to seek out challenges where failure is an option, and to value mastery goals rather than performance goals are very worthwhile things for a teacher to do.

Chapter 6: "My Best (Research-Informed) Class Ever"

◯   i.   Class periods should be designed with an understanding that what students will recall most is what takes place in the first part of the class and what students will recall second most will take place in the closing minutes of class. Use these two prime learning times well.

○   ii.   Students should be given more frequent, formative, low-stakes assessments of learning. Teachers and students should use the feedback to adjust how they teach and study next.

○   iii.   Using active retrieval methods, such as self-testing, is a more effective study strategy than using passive methods, such as rereading notes or the textbook. The latter tends to leave students with the illusion of fluency. Students need help in learning and incorporating these strategies.

○   iv.   Teaching and assessing in multiple modalities will lead to more effective learning. The modalities used should be chosen to best suit the content aiming to be delivered, not some perceived notion of "learning styles" in the class.

○   v.   Long-term memory storage means knowing something beyond the end of chapter test, or even the end of year test. Teaching with long-term memory storage in mind—which is what "learning" actually is—means that teachers need to make subtly different decisions about what they teach and how they teach it.

○   vi.   The final minutes of class are one of the prime times for deepening learning. So don't keep presenting new material right up to the bell. Instead, have an exit strategy, such as an "exit ticket," that uses the final minutes as an opportunity for the students to reflect on and recall what the instructor wanted them to learn for this class.

○   vii.   The "spacing effect" helps students store material in their long-term memory—trying to recall information, then checking to see how you did, at spaced intervals. Teachers should create these moments in their curriculum schedules, and also teach students to work like this independently. Learning is best when it is iterative.

Chapter 7: "I Love Your Amygdala!"

○   i.   Too much stress will cause "downshifting" in students' brains, and information coming in through their senses will not be processed by the higher-order thinking parts of the brain.

○ ii. Too little stress will cause disinterest. Furthermore, some degree of stress is needed for children to develop healthy stress-response systems. The teacher must, therefore, be the consummate stress balancer—knowing that the stress levels of each student at each moment is constantly shifting as the academic and nonacademic demands of the day constantly shift. The goal is to keep students in the "zone of proximal discomfort."

○ iii. Boredom invokes a physiological response that hinders learning. Teachers have a responsibility to address boredom—they cannot abdicate this fully onto students.

○ iv. Identity threat shuts down learning. Eliminating identity threat and deliberately aiding identity validation are fundamental parts of effective teaching.

○ v. Find ways to reduce or eliminate unnecessary stressors that, at their core, are barriers to learning. Remember that we often fall into the trap of confusing *creating barriers to learning* with *maintaining rigor*. Our aim is to lower the barriers, not the bar.

○ vi. Teachers should use strategies, such as choice or novelty, that help students' amygdalae choose the prefrontal cortex reflective brain path.

○ vii. Happiness and the underlying importance of positive relationships among peers and with adults in the school community help build a community of learning and learners.

Chapter 8: Memory + Attention + Engagement = Learning

○ i. Current accessibility is different from learning. If we want grades to reflect learning, we must be careful in how significantly we assign grades to measures of current accessibility.

○ ii. True learning may take a while, and we need to give students space in order to do it. Teachers must think hard about devising means of assessing the level of true learning.

○ iii. Making errors and learning from them is an important part of learning. Teachers need to create opportunities for errorful learning, and foster classroom environments where students feel safe to learn this way.

○ iv.   Teachers should introduce difficulties into learning. This reduces accessibility—students do not have the to-be-learned material at hand. When students overcome this difficulty through subsequent practice, a higher level of learning often occurs.

○ v.   Spacing, self-testing, active retrieval methods, pretesting, and formative assessment are methods that can increase learning.

○ vi.   Teaching students effective use of these methods alongside material that needs to be learned can help them be stronger independent learners.

○ vii.   Interleaving your curriculum so students revisit, recall, and build on previously learned material aids learning.

○ viii.   Teachers should provide students with opportunities for reflection on how different strategies they are using are working for them.

○ ix.   Teachers and schools have a role in educating parents in how best to support students at home.

Chapter 9: Assessment 360°

○ i.   Do your assessments, and all the preparation leading up to them, and anything you might do after them, cause your students to think hard? Are they challenging because they force, or maybe even cajole, your students to think hard? Or are they challenging because they consume lots of their time?

○ ii.   Four academic mindsets are shown to contribute to academic performance: *I belong in this academic community* (sense of belonging); *My ability and competence grow with my effort* (implicit theories of ability); *I can succeed at this* (self-efficacy); *This work has value for me* (expectancy-value theory). Teachers and schools should work to foster these mindsets.

Chapter 10: Homework, Sleep, and the Learning Brain

○ i. The benefits of homework are difficult to study. In elementary school, though, the average correlation between time spent on homework and achievement hovers around zero. Effective homework is associated with greater parental involvement and support, however. When assigning homework in elementary school, therefore, it is of prime importance to deliberately figure out the role of parental support in the context of your school and class.

○ ii. In secondary school, research suggests there is a significant difference between good homework, which increases achievement, and bad homework, which doesn't.

○ iii. Homework is most effective when used as a short and focused intervention (for example, in the form of a project or specific target connected with a particular element of learning). Benefits are likely to be much more modest if homework is more routinely set (for example, learning vocabulary or completing problem sheets every day).

○ iv. Homework is most effective when it is an integral part of the learning schema of the class, rather than an add-on.

○ v. It is important to provide high-quality and timely feedback on homework.

○ vi. It is important to make the purpose of each homework assignment clear to students.

○ vii. There is an optimum total amount of homework of between one and two hours per school day (slightly longer for older pupils), with effects diminishing as the time that students spend on homework increases. The quality of homework is more important than the quantity.

○ viii. When assigning homework, remember the vital importance of sleep in a student's schedule. Sleep is crucial for memory consolidation, and allows students to function efficiently and effectively at school.

○ ix. "Doing School" is the opposite of true learning. Make sure that the homework you assign promotes the latter, not the former.

Chapter 11: Technology and a Student's Second Brain

◯ i. In this age of technology temptations, it is important to remember the centrality of the role of building knowledge in education.

◯ ii. "The Trivium" gives an elegant way to think where technology might have the greatest impact: (1) Grammar—learn the knowledge base; (2) Dialectic—use and manipulate the knowledge base in discussion; (3) Rhetoric—communicate the results of the discussion. Of these, steps 2 and 3 are particularly ripe for the use of technology to enhance learning.

◯ iii. The essential question when exploring the value added of technology to teaching and learning is, how can this hardware or software deepen learning?

◯ iv. Technology is a part of content knowledge because our subjects have shifted, and will continue to shift, in the perhaps decades since we studied them in college. How current are we? Part of staying up to date in content knowledge is learning the technologies that are authentically germane to that discipline—including in the professional practice of that discipline and the higher-level learning of that discipline.

Chapter 12: Teachers Are Researchers

◯ i. Expert teachers need to know their content really well and must develop their content knowledge, or "curriculum understanding," alongside their pedagogical knowledge. Continued development of curriculum understanding, as well as pedagogical knowledge, should be a component of ongoing professional development.

◯ ii. Know thy impact. The successful implementation of "research-informed strategies" often relies on teachers figuring out exactly how to make them work in the context of their subject, class, age range, and school. Success is not 100 percent guaranteed each and every time. Teachers should find as many robust ways as possible to assess the impact of the strategies they try. They should also work iteratively to improve them over time.

○ iii. Whose work in your school do you respect? They may be in your subject area, but they don't have to be. Can you name two colleagues who you would like to undertake reflective, iterative, research-informed, collaborative practice with?

○ iv. Can you identify an area that your reflective, iterative, research-informed, collaborative group would like to study, such as strategies to aid memorization and recall, or the use of feedback?

○ v. Can you identify someone in your school outside of your reflective, iterative, research-informed, collaborative group who would be able to help you in your work?

○ vi. Can you identify a school leader who would support your group's work?

Chapter 13: From Research to Practice

○ i. Expert teachers identify essential representations of their subject.

○ ii. Expert teachers guide learning through highly context-dependent and situationally cognizant classroom interactions.

○ iii. Expert teachers monitor learning and provide feedback. They are adept at monitoring student problems and assessing their level of understanding and progress, and providing relevant, useful feedback. They are adept at developing and testing hypotheses about learning difficulties or instructional strategies. They do this with automatic fluency.

○ iv. Expert teachers attend to affective attributes: they have high respect for students and are passionate about teaching and learning.

○ v. Expert teachers influence student outcomes by engaging students in learning; developing in their students self-regulation, involvement in mastery learning, enhanced self-efficacy, and self-esteem as learners; providing appropriate challenging tasks and goals for students; and enhancing both surface and deep learning.

○ vi. Do you know the common neuromyths, and are you working to eliminate them from your spheres of influence at school?

○   vii.   Do you know a range of MBE research-informed strategies, and are you working to implement them in your class?

○   viii.   Are you working to spread the knowledge and use of MBE research-informed teaching and learning in your school?

Conclusion: The 10 Percent Challenge

○   i.   Will you commit to changing just 10 percent of what you do each year, trying things informed by research?

○   ii.   Have you identified any colleagues who might be interested in learning more about how research from MBE science can inform their practice, and who might join your emerging cohort of reflective, iterative, research-informed collaborators?

INDEX

In an effort to put research into practice, we wrote this book with the spacing effect in mind—we leave a gap and then come back to revisit key ideas. You will perhaps notice this as you scrutinize the index.

ABOUT THE AUTHORS

 **Dr. Ian Kelleher** grew up in Cambridge, England. He went to the University of Manchester as an undergraduate, where he received a bachelor of science degree in geochemistry. He returned to Cambridge as a PhD student at the University of Cambridge, Churchill College, working in the Department of Earth Science on the formation and stability of carbonates. After receiving his PhD, Ian moved to the United States where he spent six years teaching, coaching, and dorm-parenting at Brooks School in North Andover, Massachusetts. At St. Andrew's Episcopal School, Ian teaches chemistry, physics, and robotics, and coaches boy's JV soccer.

Ian is the head of research for the Center for Transformative Teaching and Learning (CTTL). His work focuses on the development of projects measuring the effectiveness of research-informed strategies used by St. Andrew's teachers and students. Ian also co-facilitates the CTTL's Creating Innovators through Design Thinking workshop and is responsible for the CTTL's Teacher and Student Research Fellowship Program. Ian's most recent publication is "Stress and the Learning Brain," which appeared in *Independent School* magazine in fall 2015. Follow Ian on Twitter @ijkelleher.

Glenn Whitman directs the Center for Transformative Teaching and Learning (CTTL) at St. Andrew's Episcopal School, where he also serves as the dean of studies for preschool through twelfth grade and teaches history. Glenn is a former Martin Institute for Teaching Excellence Fellow, author of *Dialogue with the Past: Engaging Students and Meeting Standards through Oral History*, and coeditor of *Think Differently and Deeply*, the national publication of the CTTL. Glenn has also written numerous articles on translating MBE research into classroom practice such as, "Assessment and the Learning Brain," which can be found in *Independent School* magazine. He is also a blogger for Edutopia. Glenn earned his MALS from Dartmouth College and his BA from Dickinson College. Follow Glenn on Twitter @gwhitmancttl.

JOIN THE NEUROTEACH NETWORK

http://www.thecttl.org/neuroteach

We know from research, and our own experiences, that reading a book on its own does not often lead to definitive changes in one's instructional practice or educational philosophy. We also know that teaching happens best in collaboration and partnership. The Neuroteach Network was launched to bring together teachers, school leaders, policy makers, and students to help solve a problem and to provide a model of how to integrate MBE research-informed teaching and learning strategies into how educators design their schools, classrooms, or work with each individual student. Sign up today!

FREE BENEFITS OF BEING PART OF THE NEUROTEACH NETWORK

- Receive *The Bridge*, the monthly Neuroteach network newsletter

FOR A ONE-TIME FEE, RECEIVE THESE ADDITIONAL BENEFITS

- Become part of an international, virtual, professional learning community and share your experiences implementing ideas from *Neuroteach* or from research you have discovered on your own.
- Reduce cost to the Center for Transformative Teaching and Learning in-person or virtual programs.
- Participate in Neuroteach Network virtual professional development hangouts.
- One complimentary copy of the Center for Transformative Teaching and Learning's publication, *Think Differently and Deeply*.

For more information about the Neuroteach Network, e-mail neuroteach@thecttl.org or follow us on Twitter @neuro_teach.